NASA Technical Memorandum 4502

I0393873

Space Life Sciences Research: The Importance of Long-Term Space Experiments

The George Washington University
Washington, D.C.

NASA Office of Life and Microgravity
Sciences and Applications
Washington, D.C.

National Aeronautics and
Space Administration

Office of Management

Scientific and Technical
Information Program

1993

TABLE OF CONTENTS

FOREWORD

This report focuses on the scientific importance of long-term space experiments for the advancement of biological science and the benefit of humankind. It includes a collection of papers that explore the scientific potential provided by the capability to manipulate organisms by removing a force that has been instrumental in the evolution and development of all organisms. Further, it provides the scientific justification for why the long-term space exposure that can be provided by a space station is essential to conduct significant research. While past history has shown that new frontiers of research offer the greatest benefits along with the greatest challenge, it does not lessen the task to justify the merits of a new and poorly understood science. Fortunately, early research results, limited as they have been, provide strong support for a vision of future scientific benefits to humankind resulting from life science research conducted in space.

We wish to thank the authors of the papers for their contributions. In addition, the technical and editorial assistance of Elizabeth Hess, Janet Powers, and April Commodore are gratefully acknowledged.

The Rationale for Fundamental Research in Space Biology: Introduction and Background

THORA W. HALSTEAD
NASA Headquarters, Washington, DC 20546

ROBERT W. KRAUSS
NASA Headquarters, Washington, DC 20546

ABSTRACT

With the construction of Space Station Freedom, NASA will have available a new platform for experiments in space that promises many advantages over those already flown. Biologists are poised to take advantage of the greater space, the increased power, and especially the long duration of the station for a cascade of innovative experiments in fundamental science that are long overdue. The unique space environment will provide new dimensions for approaching some of the most challenging problems still facing modern biology. Solutions to basic questions about living systems, which may now be grown through many generations in space, will not only explain abnormalities already observed there, but will add to our understanding of how life functions on Earth. Much will be learned about evolution that has built us the way we are, but also about what it has in store for the Earth's species in the future. NASA must not lose this opportunity to contribute to the welfare of the peoples of the Earth while at the same time create knowledge that will enable human exploration of space in the decades ahead.

INTRODUCTION AND BACKGROUND

During 1991 NASA was visited by a thorough review of its activities by the "Augustine Committee" appointed by President Bush. The committee, comprised of leaders in the aerospace industry and of persons bringing long histories of interaction with NASA, recommended some small changes in organization and directions but reinforced the view that NASA's role in the next century ought to be a major one. NASA should be committed to continue and expand its exploration of space away from Earth and also to strengthen its observation and remote monitoring of Earth itself. All that NASA could hope to do would be part of one mission or the other. With that the simplicity ended and NASA's Divisions and Centers began to reexamine their roles and organization to mesh the two missions and realign many programs already under way.

Life Sciences have shared in the general introspections. Where does space biology belong? Is it just part of medicine, or vice versa? What are the governing forces that determine the nature of the research that it funds? What can it contribute, and even, what is Space Biology?

We are not going to attempt to answer these questions, but rather examine what biological science is attempting to achieve here on Earth as well as in space. Hopefully, enough light can be generated from these papers to help NASA and the scientific community take a fresh look at biology to better plot its strategy from a different perspective. No other science is in a better position to benefit from a continuance of space platforms on which experiments can be performed, and no other science is as likely to make discoveries that will more promptly contribute to the physical and intellectual welfare of humankind. However, at this junction neither biologists, the public, NASA, or Congress are awake to biology's full potential in space in spite of over 30 years of Man's presence there. Historians in the future will surely reflect on the reasons why biology was so slow to exploit the special conditions of spaceflight to the advantage of all.

Perhaps a few words of background will set the stage for the papers that are to follow. First, it is useful to be clear in the definition of "science." Science is directed toward finding and organizing facts or data into a body of knowledge so that a given phenomenon is completely known or understood. The quest for understanding the phenomenon called life is far from over. There is now only incomplete knowledge in spite of the dramatic achievements of biologists in the last half century. Furthermore, those achievements have been characterized by a purity of purpose that is not obscured by the requirement that practical benefit must come promptly.

Second, it is well to recall that useful new technologies have flowed so consistently from basic studies that seek knowledge for its own sake that society takes those contributions for granted. That there is a role for applied research is undeniable, but even applied science advances faster when it is done in the context of sound basic facts and principles. It is therefore imperative that NASA support a basic life sciences program, as it does with physics and astronomy, to accompany an applied program in space medicine, which is essential to enable astronauts to survive and function in space for extended periods. The two programs can thrive on the synergism that is predicted upon the recognition of clear priorities for each.

Third, some appreciation of the way biology works is necessary. Although biology employs math, physics, and chemistry as its tools to elucidate living systems, the very great complexity of life and the difficulty of segregating variables in even the most focused experiment dictates a somewhat different methodology for research. Work usually begins with an exploration phase looking at performance and function, followed by an analytical phase to identify control mechanisms by repeatedly probing with carefully crafted experiments. It is initially an inductive science with broad principles evolving from a multitude of what often may appear to be unrelated experiments. Testing of principles and premises through deductive reasoning properly integrated and validated then matures into knowledge.

If it is expected to yield sound knowledge of enduring value biology requires time to observe, experiment, replicate, and deduce. It demands increasing experimental sophistication as it creates and organizes data into the fabric of truth. In space, life science will require the same dedication and continuity of a large research cadre that has characterized its profound scientific advancements on Earth.

Finally, it is useful to recall that biology in space is no different from that on Earth, because anything living we will carry from Earth into space. Biological science is not looking for new space entities — black holes, quasars, galaxies, quarks, and the like — as the physicists are. Biologists will deal with how species that evolved on Earth react to the space environment. By understanding and being able to predict the performance of organisms, tissues, enzymes, and genetic codes in space, biologists can test and refine their science. They can learn about the potential of protoplasm to exist, function, and organize in forms not constrained by gravity, though confined and interdicted by the unique radiation spectrum found there. The living horizon can prove as unlimited as the edge of the universe.

In plotting the future course for NASA's Life Sciences the fundamental scientific questions are preeminent. They deal with (1) how genetic codes turn on and off, (2) how totipotent cells evolve specialized tissue, (3) why cells suddenly revert to totipotency and go out of control, (4) what controls aging, (5) how evolution operates, and (6) why, with current knowledge of genetics, genes, protein biosynthesis, and enzymatic control, has life not yet been created *de novo in vitro* in the laboratory. There are many others that experience and experiments in space can address from a new vantage point.

The environment in orbits or on long-term interplanetary flights is characterized by:

1. A reduction in the force of gravity to near zero.
2. A space radiation spectrum.
3. A sealed and confined environment.
4. Combinations of all of these.

The task facing the biologist who aspires to contribute to science through space experimentation is to employ these parameters in ways that provide new perspectives to the old problems.

NASA's space science must interact and integrate with the main scientific thrusts of modern biology if it is to bring the unique tools that space affords to answer problems common to all. The new information that will be developed about how organisms cope with space and the new biochemistry revealed will be of immense practical value to those who hope to endure in space, but the scientific target must not be obscured.

In space, as on Earth, the greatest rewards to biological science can be expected to come from understanding, predicting, and ultimately controlling the process and progress of specific genes in organisms and their phenotypic manifestations. The development of such knowledge can be vital to the welfare of humankind in the centuries ahead. It will not be created from studying response to gravity alone. However, no single physical force has been so pervasive, so constant, and so unavoidable as gravity during the millions of years of evolving life in light or darkness, in heat or cold, in water or in air, and at all pressures and elevations. The new opportunities to probe the responses to such a profound determinant of all life on Earth are too precious to ignore any longer.

Opportunities and Questions for the Fundamental Biological Sciences in Space

JOSEPH C. SHARP
NASA, Ames Research Center, Moffett Field, CA 94035

JOAN VERNIKOS
NASA, Ames Research Center, Moffett Field, CA 94035

ABSTRACT

With the advent of sophisticated space facilities we discuss the overall nature of some biological questions that can be addressed. We point out the need for broad participation by the biological community, the necessary facilities, and some unique requirements.

INTRODUCTION

Fundamental biological science has many meanings to many people. To us, it means the pursuit of new knowledge about life. With the advent of the space program and the writing of the NASA's charter, access to space suddenly made it possible for the nation's scientists to explore a previously unavailable, but critical (perhaps the most critical) element in the evolution of living systems on Earth. Life has evolved on Earth in the continuous presence of gravity. By going into space, we have been given a glimpse of how gravity may influence biological systems. However, even Soviet achievements of one year in Earth orbit are but a minute fraction of the hundreds of millions of years it has taken life to evolve on Earth. After more than three decades, new knowledge about the physical nature of our solar system, our galaxy, and the cosmos has stimulated, fascinated, thrilled, and dominated scientists involved in the space program, the general public, the young, and the old. On the other hand well over half of NASA's budget has been tied to the manned program and its support. The life sciences program has largely focused on the medical and operational aspects of flying people in space. The basic biological sciences have received but desultory and sporadic attention and support. Why did physical sciences succeed where the biological sciences has lagged so far behind? We are not going to discuss the reasons for the relative paucity of work utilizing the space environment by the community of biological scientists. Rather, we shall focus on the opportunities we see to excite the scientific community into participation.

Unlike the physical sciences, biology is frequently a long, complex, interactive process, not a singular event. Therefore, it requires a continuum of formulation of hypothesis, manipulation, interpretation, reformulation, and replication, which necessitates repeated access to the variables examined. In practical terms, what that means is that in biological sciences, a single flight experiment serves only to whet the appetite and to more accurately point the directions for the next experiment. In other words, a single mission will not answer a biological question. In order to acquire new knowledge in gravitational biology, we need to make this message clearer to NASA and to the public and develop ways to enable long-term commitments between the scientific community and NASA.

The study of biology in space can be divided into four major categories: (a) the study of the origin of life, its distribution, and fate; (b) the utilization of the perspective from space to better understand the complex interactions between the biological and physical, global properties of Earth; (c) the specific exploitation of the microgravity environment to better understand the fundamental properties of life here on Earth; (d) the utilization of spaceflight as a unique form of provocative stimulation to better understand the mechanisms by which living systems respond and adapt. This last approach addresses most closely the acquisition of knowledge in support of space medicine and manned flight. However, it is the pursuit of new knowledge of the effects of gravity and microgravity on living systems that we will dwell on here. Our approach will be to ask and then discuss some first and second order questions about basic issues in biology. These issues include: structure, from cell to organ to organism; function, the regulation of systems such as immunology, neural sciences, and behavior; and reproduction and development. We have selected questions from each of these classical subdivisions of gravitational biology in order to show the depth and profound diversity of disciplines that could actively participate in this relatively new science.

STRUCTURE

Gravity can be envisioned as a load generating physical stress, and as a sensory input. How do individual cells and populations of cells perceive gravity?

Do cells directly perceive and respond to gravity or is gravity's influence upon cells mediated indirectly? Cells have a structural system — the cytoskeleton — that provides them with shape and dimension. In addition to its structural features, the cytoskeleton acts as a sensory organelle. Cells make mechanical connections to the substrate they grow on, to neighboring cells, and to soluble circulating factors like growth factors. These connections can be relayed by the cytoskeleton. They can also be relayed by internal chemical pathways that transmit by cascade action signals that are sensed at the cell membrane and that need to be sent to the cell nucleus where the genetic material, DNA, resides.

From an experimental perspective, what information (stimuli) is passed to a cell living in a three-dimensional body as opposed to one living on a two-dimensional, *in vitro*, cell culture? At the cellular level and in a microgravity environment, what is the relationship between function and structure? If there is a relationship at this level of analysis how does it affect cell differentiation, development, carcinogenesis, or cell senescence?

Autonomous, individual cells suggest other questions. These independent units provide all the functions necessary to life in one compact package. They evolved long before multicellular organisms, like us, with our specialized cell functions. Questions of the role of gravity in evolution, and the adaptability of terrestrial life to altered gravities, surround the study of these types of biological systems.

To go a few steps higher in the living system and its integration: Is gravity necessary for the normal development of a musculoskeletal system? How does the presence or absence of gravity influence the deposition of mineral in matrix? What are the systemic mechanisms involved in the adaptation from Earth gravity to the microgravity of spaceflight? What are the gravity thresholds for proper organ and system development? Do the usual risk factors such as gender, age, nutrition, exercise, species, or strain strongly interact with exposure to altered gravitational forces? How do they interact with the radiations found in deep space or other spaceflight associated factors?

FUNCTION

In animals as well as higher plants there are systems that respond to the acceleration force called gravity. How do they adapt to altered gravitational environments? Will organisms that mature in microgravity or altered gravitational environments develop phenotypically or functionally different gravity sensors? What are the thresholds and functional dependencies of acceleration for the various gravity sensors? How can the rapid adaptation to microgravity by animals, including humans, be used to understand the plastic nature of the nervous system? Earth's gravity is such that organisms raised on Earth seem to adapt reasonably well to microgravity; is the reverse true?

What is the consequence of altered gravity perception on musculoskeletal function? On smooth muscle? How are these translated into altered motor behavior and autonomic functions? Similarly, what is the consequence of altered perception on endocrine feedback regulating systems? How is that translated into altered metabolism, arousal, sensory thresholds, and reproduction? What is the consequence of altered perception on the regulation of circadian rhythmicity? Is gravity a major *zeitgeber*?

REPLICATION, REPRODUCTION, AND DEVELOPMENT

Mammalian cells, in certain growth stages, present particular questions with respect to gravity-mediated effects. During fetal development, spatial orientations and associations with certain substrates are critical in the proper execution of programmed development, differentiation, and growth. This occurs in the buoyant environment of the womb akin to the marine environment where life evolved, and where gravity is perceived quite differently than after birth. The activated, dividing behavior of fetal cells is partially recapitulated in adult life during the processes of healing and repair and, in the case of cancer cells, where normal growth controls are bypassed. The relationship between structural forces provided by or enforced by gravity in these growth states can be determined only by studying the effects of altered gravity on these processes (Grymes, 1991).

Mammalian reproduction and reproductive behavior are particularly sensitive to perturbations. It will be particularly challenging to isolate the effects of gravity on these functions, since numerous spaceflight associated conditions are well known to interfere with this process. However controlled these experiments might be, it may not be for several generations, following adequate adaptation to these conditions, that the true effects of microgravity on reproduction and reproductive behavior may become evident.

In both plant and animal systems, the concept of critical periods in development, wherein experimental intervention can irreversibly alter neural circuitry, adult sexual behavior, or endocrine responses (Vernikos, 1972), suggests that gravity may also exert its most profound effects at these times.

Furthermore, the effects of gravity and microgravity on life span as well as seed-to-seed and generation-to-generation morphological and functional evolution need to be addressed.

THE NEED FOR CONTROLS

Biological research is complex since, at all levels, from a single cell to entire organisms, there are so many interacting, mutually dependent subsystems. Such research depends heavily on the elimination of interfering variables by conducting appropriate controls. This is particularly and overwhelmingly true in the spaceflight scenario. Delay between loading and experiment and inflight access, lift-off forces, need for remotely controlled manipulation, re-entry forces, and delay in accessing experimental specimens postflight are all unique and difficult to control variables. For example, microgravity-induced alterations of the immune system have been reported. However, inflight samples from animal species have yet to be obtained or analyzed. It is now increasingly evident that acute immune responses can be measured in humans following postural change or exercise so that differences in pre- and postflight data may be accounted for by re-entry and landing events. Similarly, inflight human evidence may be due to microgravity or to the confinement of spaceflight or some other environmental variable, which, so far, has not been controlled.

The requirement for an onboard centrifuge to provide a 1 g simultaneous control could reduce much of the ambiguity present in many past studies. Of course, centrifugation may well introduce new and unexpected variables. The validity of a conclusion that a particular biological phenomenon is, indeed, due to gravity or its absence is one not only where all possible other explanations have been systematically eliminated, but also where the phenomenon can be demonstrated in multiple species, including humans.

FACILITIES

What do we need to conduct such research? At the very least, continuity and the ability to conduct repeated experiments in the same laboratory are required. The Soviet Cosmos unmanned biological satellite program, which launched multispecies experiments approximately every two years since 1972, has proven the value of such an approach. As we become more sophisticated in the use of artificial intelligence for inflight, remotely controlled manipulation of payloads, an unmanned, recoverable, free-flying untended platform that exposes specimens to prolonged periods of microgravity (e.g., greater than 60 days) could form the bread and butter of a biological sciences program. It is clear, however, that such an unmanned satellite could never replace the need for a human-tended, permanent, Earth orbiting laboratory. Such a laboratory should make it possible to study, on orbit, significant numbers and varieties of experimental specimens, with appropriate 1 g controls and the capability for observation, intervention, and testing. It does not have to be elaborate, but it is essential if gravitational biology is to move forward — away from simple parametric observations.

CONCLUDING REMARKS

The history of biological science (as well as all science, for that matter) is replete with examples of discovering deep and profound new knowledge upon gaining control of a primary physical variable, *viz.* light, momentum, sound, chemistry, and radiation, to mention but a few. There is every expectation the same will be true for gravity since now, for the first time in history, we can "control" or manipulate accelerations to less than the equivalent of 1 g.

The questions raised in this paper are but a few examples. It is up to the biological scientific community to harness their creativity towards this exciting research frontier. The facilities to conduct the research are expensive and complicated, yet some are already available to our nation's scientists; better ones will become available in the not-too-distant future. Support for biological research in space will happen only if the scientific community strongly believes, as we do, in its value and potential. Together we can capture the imagination of the public and persuade them of the benefits. The laboratories in space will always be a scarce and expensive commodity; we must make sure that as scientists we are selective and apply the highest scientific rigor to experimental design and data interpretation. On our part, we at NASA must develop a way to simplify procedures for enabling science to be conducted in space. A broad foundation of ground research, addressing specifically these questions, needs to be developed and nurtured before the jump to flight is made.

Ground and flight scientific programs are inexorably intertwined and although ground facilities exist, the community to support a space laboratory is inadequately small.

The reality of experimental control of gravity is within the reach of biologists; using this opportunity properly, we will reap new and exciting insights into life. With such insights we will be able to make intelligent and efficient advances as humankind continues to personally explore the limitless frontiers of space. We can only speculate about findings that will permeate our understanding of Earthly biology ... life as we know it. The entire history of science indicates it is certain that the new knowledge will be important to furthering our understanding of biology: our personal origins and fate!

REFERENCES

Grymes, R. 1991. Space Biology Teachers' Conference, NASA Select TV, December 11, 1991.

Vernikos, J. 1972. Effect of hormones on the central nervous system. In: *Hormones and Behavior* (Levine, S., Ed.), p. 11-62. New York: Academic Press.

Space Research with Intact Organisms: The Role of Space Station Freedom

ROBERT W. PHILLIPS

NASA Headquarters, Washington, DC 20546

FRANCIS J. HADDY

Uniformed Services University of the Health Sciences, Bethesda, MD 20814

ABSTRACT

The study of intact organisms has provided biologists with a good working knowledge of most of the common organisms that have evolved in the 1 g environment of Earth. Reasonably accurate predictions can be made about organismal responses to most stimuli on Earth. To extend this knowledge to life without gravity, we must have access to the space environment for prolonged periods. Space Station Freedom will provide a facility with which to begin this type of research. Spaceflight research to date has been limited to relatively short-term exposures that have been informative but incomplete. This paper provides a brief background of known changes that have occurred in intact organisms in the space environment and proposes the kinds of experiments that are needed to expand our knowledge of life on Earth and in space.

INTRODUCTION

The greatest challenges and the greatest opportunities for space and gravitational life science research will come with the study of intact organisms. Such research will utilize many species, from simple prokaryotes through humans. They, with their multiple systems, have evolved over countless generations under the constant influence of the Earth's gravitational field into the familiar plants and animals that we recognize today. These organisms have been studied, analyzed, and dissected functionally, morphologically, and chemically by today's scientists and their predecessors. Most biologists would agree that we have a good working knowledge of most of the common organisms in our environment, at least at the organ and system levels. In recent decades increasingly sophisticated research tools have allowed scientists to probe more deeply into biological function. These efforts have begun to provide an understanding of basic mechanisms at the cellular and subcellular levels of organization. Reasonably accurate predictions can be made about animal responses to most stimuli on Earth. No such storehouse of knowledge exists concerning organismic response to the stimuli found in space. Only within the past few years has there been the opportunity to study organisms exposed to the space environment, removed from an absolute environmental constant, "the force of gravity." The evolution of all life has occurred in the 1 g environment that our bodies recognize as the norm.

Viewed from another perspective, the law of gravity is the one law that cannot be broken, modified, or ignored as long as we continue to live on the face of the Earth. An excellent analogy to the problem of trying to study the effects of gravity while restricted to ground-based facilities was suggested by A.H. Brown. Imagine a student of the effects of light being unable to utilize darkness as a test paradigm. The student might modify the position of the light, or make it brighter (hypergravity), but could only turn it off for brief instants (free fall). Without the ability to investigate the role of darkness for prolonged periods, could the real roles of light with all their subtleties ever be established? For that reason the opportunity to examine the behavior and function of organisms removed from their hereditary gravitational environment is unique.

To date that opportunity has been more promise than fulfillment. There have been a number of preliminary descriptive reports of the immediate, short-term responses during and following exposure to the weightless and the combined weightless and high radiation environments of space. These studies have been informative and in many cases intriguing. Unfortunately, they leave many questions. In almost no instance have adaptive responses been carried to new stable endpoints. Developmental biology and multiple generational studies are still dreams awaiting the availability of long-term laboratories in space.

Space Station Freedom, even with its diminished capacity following restructuring, will provide a facility in which such studies can begin to be made. What could and should be studied? How can biologists most effectively utilize the life science research facilities on Freedom? The intent of this paper is to provide a brief background of the changes that have been noted in intact organisms exposed to space and suggest some examples of the kind of experiments that might provide new and exciting information on the role of gravity in the evolution of life as we know it and how gravity has shaped function and morphology in every intact organism — terrestrial and aquatic.

THE EFFECTS OF SPACEFLIGHT

Humans

The intact organism that has been studied most often in both the American and Soviet, now Russian, programs is the human being. By and large these studies have dealt with problems of immediate concern to operational medicine. Certainly we can now recognize and even anticipate a number of acute and semi-chronic adaptive responses to microgravity. The most prominent changes can be directly related to the weightlessness that is a characteristic of spaceflight. Other changes may be due to anxiety, changes in activity, or generalized stress. It should be emphasized that many of the changes that occur are truly adaptations to a novel environment and are appropriate as long as one remains in space. There are no obvious major "in-flight" detrimental manifestations once the very acute alterations of the first few days, such as space motion sickness, subside. Real and potential problems become evident following return to Earth's gravity field.

Spaceflight produces many changes in the human body. Some are minor and both develop and subside in the first few days, such as motion sickness, which is present in about 60% of space travelers. Facial edema, decreased red cell mass, and a transient neutrophilia are also components of the early response to microgravity. The body's immediate responses also include shifts in fluid from the dependent portions of the body, as well as decreases in the size of the various water pools, including blood volume. The fluid shift to the upper body begins to occur in humans as they lie in a leg-elevated position prior to launch. The shift results in a condition that has been colloquially called "bird legs" because the shift greatly reduces leg girth.

There are changes in the cardiovascular system that are often described as deconditioning. In actuality, the appearance of deconditioning becomes apparent primarily following return to Earth, and is characterized by decreases in stroke volume, blood pressure, and an increase in heart rate.

Some of the changes that occur during spaceflight are more serious and occur more slowly, such as skeletal muscle atrophy, bone demineralization, and psychosocial problems. An increased potential for radiation damage is superimposed when the flight is into deep space instead of low-Earth orbit. Skeletal muscle and bone atrophy represent major long-term

adaptive responses that most investigators feel are analogous to disuse atrophy on Earth. It has been determined that the most prominent muscle changes are in the slow-twitch antigravity muscles, which are tonically active on Earth but not required for maintenance of posture while in space. Morphologically, individual muscle fibers are diminished in size. Functionally, based on enzyme concentrations, there is a switch from slow-twitch to fast-twitch fiber types. Similarly, the lack of weight-bearing stresses on the skeletal system in space decreases the need for large, dense bones. Massive remodeling of the skeleton is initiated with calcium mobilization dominant over calcium deposition. The result is an osteoporosis-like decrease in bone mass, which may be continued well beyond the one year that Soviet cosmonauts have spent in space. Overall, the rate of calcium loss from the body in humans is of the order of 1% per month. However, the loss is not uniform in all parts of the skeleton, and the complex changes may affect structure more than mass.

There are also alterations in the neurovestibular system. The most notable alteration occurs in the first few days of flight as a malaise that is variously called space adaptation syndrome, or more explicitly, space motion sickness. It is a transient response seen in over half of all space travelers. Although the specific etiology is still open to debate, there is a reasonable consensus that it is related to sensory conflicts between the visual and vestibular systems with additional central nervous system modification of the activity of the autonomic nervous system. Of greater consequence are more chronic central nervous system changes that do not appear to be manifested while inflight but become prominent following return to Earth. These include both sensory and motor effects such as altered balance and hand-eye coordination. A good review of the effects of spaceflight on physiological systems has been presented in the recent book by Nicogossian et al. (1989). Additional information is available in the Proceedings of the Space Life Sciences Symposium (1987).

Given the good hindsight present in most of us, many of these responses now seem eminently predictable. However, prior to spaceflight most of these changes were not particularly anticipated. With that as a background, how well are we able to foretell the responses that are likely to be seen in humans and other mammals maintained for prolonged periods in space or in reduced gravitational fields?

Other Animals

Although there are more data available on humans than on other organisms, there have been some studies conducted with plant and other animal species. By and large data collected from mammalian vertebrates, such as non-human primates and rodents, indicate that their changes are similar to the responses seen in humans. Certainly there are differences in magnitude, but the basic adaptations are analogous. Bone loss, muscle atrophy, and cardiovascular and sensory-motor changes are evident following return to Earth. To date, inflight measurements, other than observational, have not been made on animals.

Much of the flight data on other organisms, although tantalizing, is fragmentary. Certain simple studies must be repeated or extended for longer times. It is not the intent of this paper to present a broad review of past and current space research on intact organisms, but rather to cite some examples as a prelude to defining our thoughts on where organismic space research is needed as the opportunity develops to utilize the facilities of the space station.

A study of the effects of five days of spaceflight on avian embryogenesis demonstrated that two-day-old embryos did not survive, although they continued to grow for the first day or two following launch. Conversely, nine-day-old chick embryos were capable of continuing their development and were ultimately hatched following return to Earth (Vellinger and Deuser, 1990). Calcium mobilization from the shell was not impaired in the older embryos and their growth following hatching appeared normal (Hester et al., 1990). However, they had a decreased vestibular response to gravitational stimuli (Jones et al., 1990).

In a preliminary experiment, it was found that encysted brine shrimp (Artemia) embryos, when activated in space, grew and developed normally for the rest of the flight. Hatching and survival rates were comparable to ground-based controls (DeBell et al., 1991). Other invertebrates also appear able to develop in space. Jellyfish (Aurelia) polyps, when activated during spaceflight with iodine or thyroxin, undergo metamorphosis to produce free-swimming ephyrae that appear normal (Spangenberg, personal communication). Further, it has been reported that paramecia multiply more rapidly in space than do ground controls (Richoilley et al., 1986). Based on the responses of these very diverse invertebrate species, it would appear that aquatic invertebrate development during a single generation is not adversely affected by microgravity. Conversely, invertebrate aging and longevity were detrimentally affected by spaceflight in a terrestrial organism, the common housefly (Musca domestica). The flight animals had a greater rate of mortality and an increase of brain lipofuschin (Marshall et al., 1990). Increased brain lipofuschin concentration is associated with aging in humans.

Plants

Plant growth is also affected by the microgravity of space. Most reports have indicated that development halts at or just before flowering. In general, both root and shoot growth has been less than seen in ground controls (Halstead and Dutcher, 1987). In only one instance have plants (Arabidopsis) been carried throughout a complete reproductive cycle with flowering and seed development (Merkys and Laurinavichius, 1983). Root growth in the absence of a guiding gravity vector becomes random, and no longer orients toward ground water and nutrients. A unique, recent report states that root growth is markedly enhanced during spaceflight with little influence on shoot growth (Levine and Krikorian, 1991).

Chromosomal aberrations are also more common in plants grown in space. Basic biochemical changes have been recorded. A number of researchers have noted decreases in starch-containing amyloplasts as well as the cell wall constituents, cellulose and lignin (Halstead and Dutcher, 1987). Corn and mustard spinach seedlings exhibited a decrease in the amount of starch in amyloplasts, with an increase in the number of lipid vacuoles. Fatty acid metabolism was also modified with a decrease in the C-18 unsaturated fatty acids and an increase in the C-16 saturated fatty acid (palmitic), which is more typically a component of animal fat (Lewis and Moore, 1990).

THE FUTURE — LONG-TERM EXPOSURE TO SPACE

The exploration of space is, and should be, a transitional, stepwise process. We must walk before we run and we must float a little in low-Earth orbit before we cast ourselves on the ocean of interplanetary space. As noted above, our knowledge of the effects of space on biological function is not only rudimentary and fragmentary, it is also, with only a few exceptions, based on very short-term exposures. In these brief excursions there has been little to indicate that adaptations have reached stable new set points. In many cases the assumption has been made that acclimation is complete, but that is more conjecture than fact.

Several important questions must be addressed concerning the effects of the space environment, both the lack of a gravitational stimulus and the presence of increased quantities of a unique radiation, on intact organisms. The first question involves the gravitational stimulus in a single life cycle. Here there is a distinct difference between plant and animal kingdoms. It has been repeatedly shown that germination and early plant growth are not greatly affected by the space environment, whereas maturation, flowering, and seed production are clearly inhibited. What is not known is to what extent that inhibition is due to environmental factors other than gravity. Habitats with poorly engineered provisions for optimal plant growth and inadequate monitoring equipment will fail to expose the real effects of gravity. NASA needs to work closely with the plant science community to develop sealed habitats that will effectively isolate the gravity variable. For instance, light intensity, spectra, and duration are all important variables that must be measured and controlled. Plant hormones and byproducts such as ethylene have not been measured, due in part to resource limitations, but their lack or excess may significantly modify plant growth characteristics and the completion of maturation with viable seed formation.

In the animal kingdom, not only is there scant data on fertilization, differentiation, and embryogenesis, but later events in the developmental life cycles are unknown. In part the discrepancy is due to the difference in generation time. Only a few invertebrate animals have sufficiently short life cycles to allow generational studies with our current spaceflight systems. To date there are no data to support or refute the hypothesis that a vertebrate animal can come to sexual maturity and reproduce in space. We do not know whether gametogenesis will occur, the estrus cycle will be initiated, fertilization will take place, and, in the case of mammals, that gestation, parturition, and lactation will be normal. The one data point that we have on early avian embryogenesis indicates complete failure; all of the two-day-old embryos died within the first 48 hours. Conversely, amphibian embryogenesis was successful. *Xenopus* eggs fertilized in space developed into tadpoles,which subsequently underwent metamorphosis following return to Earth (Souza, personal communication).

Assuming that mammalian reproduction is possible through parturition, there will be other logistic problems associated with non-primate postnatal development. Imagine a litter of mice, or rats, or pigs, or puppies born in space. What kind of nest must we devise to allow the female access to her young for nursing? A unit or facility must be small enough to retain the young, yet allow the dam to enter for nursing and social interaction, and then leave to acquire food and water and eliminate body wastes, and still prevent the neonates from floating off. Such a unit will be a challenge to develop. Will the young be able to seek, find, and attach to the mammary gland to gain nutrition and the psychosocial interactions necessary for later life? Will the lack of the communal relationships of a traditional nest and the modification of early neonatal imprinting impair them as adults in their interaction with others of their species, as well as with humans?

The phenotypic changes seen on exposure to space are similar in plants and terrestrial vertebrates. There is a decrease in those morphological elements that are required to sustain the organism in the Earth's gravitational field. Practically and philosophically there is no difference between a decrease in cell wall lignin and cellulose, and a decrease in bone mass and atrophy of the antigravity muscles. Certainly the mechanisms are unique, but the fundamental changes are the same. Without a gravitational stimulus there is a decreased requirement for the structures that organisms have developed to support themselves on Earth and stand against the Earth's gravitational pull. What about aqueous organisms? We do not refer to the benthic animals that must support themselves against a gravitational field on the ocean's floor, but rather to the neutrally buoyant organisms that are free swimming. Would their morphologic development in space be modified? Is there a basis for suggesting that trout or shrimp depend on other than the resistance of their environment for bone and muscle development?

Today we cannot even say what the phenotypic expression will be in a terrestrial vertebrate conceived and grown to maturity in space. There is no predictive basis for describing the morphological changes that will occur. Bone and muscle mass will be diminished, of that there is no question, but relative changes in different parts of the skeleton and alterations in total skeletal muscle are conjectural.

Of even greater interest and concern is the generational stability of phenotypic expression. How stable and invariant is the gene pool when intact organisms are continually exposed to a new environment? For how many generations will adaptive change continue to occur? While F_1, F_2, and F_3 generations will express different phenotypes as they mature, will the changes be apparent at birth? How rapidly will individual adaptation be translated into a new genetic stability? This is an important question for both the plant and animal kingdoms. Will there be a difference? Are either plants or animals inherently more adaptable to dramatic climatic changes such as the removal of an heretofore environmental constant?

An even more basic question is, how adaptable are we? Certainly intact organisms inhabit almost all regions of the Earth, including many that at one time seemed too inhospitable for survival. The human race, in its development, has spread over most of the Earth's surface, thriving from Arctic to tropics, from mountain to lowland, from desert to rain forest. Based on our ubiquitous presence we could be commended on our adaptability. These adaptations, however, have occurred over countless generations. The very basic question is, can we or other gravity-developed organisms survive and adapt in a weightless environment? Have intact organisms in general become too specialized, too dependent on gravity, to exist and conform to a zero gravity life?

There are two major reasons why the study of humans on Space Station Freedom will not provide the answers to basic questions of adaptation. First, there is the question of time. Time as a factor in chronic adaptation of even an individual on the space station is extremely limiting, to say nothing of generational effects. The nominal crew stay on orbit following permanent manning of the facility will be 90 days, perhaps extending to 180 days — essentially one half a year. We would hold that the adaptation of a mature adult human over such a brief period of his/her life span will not answer basic questions of adaptation, nor of our ability to adapt to an essentially gravity-free life. A life span of 90 years is not at all uncommon in today's world, and a small six-month segment will not provide definitive answers to the question of long-term adaptation. It is reasonably clear from the few Soviet cosmonaut exposures of longer than six months that adaptive end-points were not present in some measured systems such as bone.

The second reason deals with the fact that many of the recognized rapid adaptations to space living are seen as detrimental upon return to Earth. Sufficiently detrimental that a major program is being instituted — Biomedical Monitoring and Countermeasures — to insure that on Space Station Freedom humans do not adapt in ways that might prove detrimental to their subsequent life following spaceflight. To the extent that this program is effective, space adaptation will be not only reduced, but prevented in the human population. To understand chronic multigenerational effects of spaceflight, it will be necessary to utilize smaller animals with rapid reproductive cycles as models of the likely responses in our species.

Questions such as those posed in the preceding paragraphs need to be addressed. We need to establish research goals that will provide fundamental information on how gravity has and does shape life on Earth. A first step in providing some answers will come from utilization of the life science research facilities on Space Station Freedom. That facility will provide a beginning in the quest for basic information on the role of gravity in the development of life. It is, however, the next logical step.

As the space station is currently configured, in the assembly phase, which includes man-tended capability crews that will be present for limited periods while the space shuttle is present, the major research emphasis will be on materials research rather than life sciences. As permanent manned capability

is developed, the Space Station Freedom program will have a gravitational biology facility, and the centrifuge facility will be added with plant and animal habitats. With these components in place it will be possible to conduct experiments leading to answers to some of the biological questions raised above. The centrifuge is designed to provide long-term exposure to 1 g fields as a control, to condition plants and animals to the force of gravity prior to initiating experiments, and also to have the capability of exposing plants and small animals to variable g forces as might be encountered on the moon or Mars.

In conclusion, we have a rare opportunity. Simultaneously we can begin to exploit the space frontier and enhance our basic knowledge of life here on Earth. The ability to conduct long-term experiments with intact plants and animals, and to have a centrifuge for providing 1 g controls and for studying gravitational thresholds, will provide important new insights. Results emanating from such work will be used in countless applications which cannot be predicted at this time. Such has always been the course of major new enterprises.

REFERENCES

DeBell, L., Spooner, B.S., and Rosowski, J.R. 1991. Reinitiation of brine shrimp embryonic development from gastrula-arrested dormancy during Space Shuttle flight. *ASGSB Bulletin* 5: 58.

Halstead, T.W. and Dutcher, F.R. 1987. Plants in space. *Annual Review of Plant Physiology* 38: 317-345.

Hester, P.Y., McGinnis, M.E., Vellinger, J.C., and Deuser, M.S. 1990. Avian embryogenesis in microgravity aboard Shuttle STS-29: Effect on shell mineral content and post g-hatch performance. *ASGSB Bulletin* 4: 25.

Jones, T.A., Vellinger, J., Hester, P.Y., and Fermin, C. 1990. Effects of weightlessness on vestibular development: Evidence for persistent vestibular threshold shifts in chicks incubated in space. *ASGSB Bulletin* 4: 75.

Levine, H.G. and Krikorian, A.D. 1991. Shoot growth, root formation and chromosome damage results from the Chromex I Experiment (Shuttle Mission STS-29). *ASGSB Bulletin* 5: 28.

Lewis, M.L. and Moore, R. 1990. Altered fatty acid metabolism in seedlings germinated in microgravity. *ASGSB Bulletin* 4: 77.

Marshall, G.J., Fras, I.A., Kirchen, M.E., Gruber, H.E., Sweeney, J.R., and Stover, S.J. 1990. Effects of microgravity on aging in the housefly. *ASGSB Bulletin* 4: 26.

Merkys, A.J. and Laurinavichius, R.S. 1983. Complete cycle of individual development of *Arabidopsis thaliana* Heynh. plants on board the "Salyut-7" orbital station. *Doklady Akademii Nauk SSSR* 271: 509-512.

Nicogossian, A.E., Huntoon, C.L., and Pool, S.L. (Eds.) 1989. *Space Physiology and Medicine*, 2nd Ed. Philadelphia: Lea and Febiger.

Richoilley, G., Tixador, R., Gasset, G., Templier, J., and Planel, H. 1986. Preliminary results of the "Paramecium" experiment. *Naturwissenschaften* 73: 404-406.

Souza, K. 1993. Personal communication.

Space Life Sciences Symposium: Three Decades of Life Science Research in Space. 1987. Proceedings of the Symposium held June 21-26, 1987, in Washington, DC.

Spangenberg, D.B. 1991. Personal communication.

Vellinger, J.C. and Deuser, M.S. 1990. Avian embryogenesis in microgravity aboard Shuttle STS-29: Experimental protocol and results. *ASGSB Bulletin* 4: 74.

Space Research on Organs and Tissues

MARC E. TISCHLER
Department of Biochemistry, University of Arizona, Tucson, AZ 85724

EMILY MOREY-HOLTON
NASA, Ames Research Center, Moffett Field, CA 94035

ABSTRACT

Studies in space on various physiological systems have and will continue to provide valuable information on how they adapt to reduced gravitational conditions, and how living in a 1 g (gravity) environment has guided their development. Muscle and bone are the most notable tissues that respond to unweighting caused by lack of gravity. The function of specific muscles and bones relates directly to mechanical loading, so that removal of "normal forces" in space, or in bedridden patients, causes dramatic loss of tissue mass. The cardiovascular system is also markedly affected by reduced gravity. Adaptation includes decreased blood flow to the lower extremities, thus decreasing the heart output requirement. Return to 1 g is associated with a period of reconditioning due to the deconditioning that occurs in space. Changes in the cardiovascular system are also related to responses of the kidney and certain endocrine (hormone-producing) organs. Changes in respiratory function may also occur, suggesting an effect on the lungs, though this adaptation is poorly understood. The neurovestibular system, including the brain and organs of the inner ear, must adapt to the disorientation caused by lack of gravity. Preliminary findings have been reported for liver. Additionally, endocrine organs responsible for release of hormones such as insulin, growth hormone, glucocorticoids, and thyroid hormone may respond to space-flight.

INTRODUCTION

Microgravity, the decreased effectiveness of gravity, produces profound effects on the body's biochemistry and physiology. Systems affected include muscle, bone, cardiovascular, pulmonary, neurovestibular, liver, and endocrine. For humans to live and work in space, it is essential that we identify the precise consequences of exposure to microgravity and then develop appropriate countermeasures, if necessary. Furthermore, understanding how animals and humans adapt to microgravity will provide a clearer picture of how gravity has influenced the development of these systems under 1 g conditions. The key word here is "adaptation." Microgravity should not be considered a pathological state. Organs and tissues are simply adapting to the new physi-ological state or environment, just as one's body must adapt to a change from functioning at sea level to performing at high altitude. For instance, decreased oxygen pressure at high altitude requires an increase in the number of red cells in the blood for transporting vital oxygen to peripheral tissues and organs. Thus the cardiovascular system is exquisitely sensitive to changes in the physiological environment of the organism.

Our precise knowledge of the effects of microgravity on humans and animals is still limited for several reasons: (1) the sample size is too small to make many generalizations; (2) the capabilities for scientific studies, including long-term flights, is limited; (3) astronauts have used a variety of countermeasures, thus obscuring effects of weightlessness; (4) there is considerable variation in the types of missions flown including duration, numbers of crewmembers, and goals of the mission, including the planned animal studies; (5) a difficulty with interpreting flight data is the tremendous variability among human subjects, which may be related to their physical status at the onset of flight; and (6) our space program faced a major setback because of the Challenger disaster in 1986. Therefore, most studies have concerned animals using primarily Earth-bound models to mimic the potential effects of unloading, the removal of the weight or mass supported by muscle or bone using artificial support or microgravity. These model systems are characterized by hypodynamia, the deprivation of normal weight-bearing function, and hypokinesia, the deprivation of normal locomotive function, both of which are common to space travel. It has also been possible using these models to mimic the body fluid shifts expected in spaceflight, as a means of testing alterations of the cardiovascular system. However, some alterations of organs and tissues caused by microgravity are not reproducible in Earth-bound animal or human models. Thus, space research on organs and tissues is essential both for validating the Earth-bound models used in laboratories, and for studying the adaptations to weightlessness that cannot be mimicked on Earth.

MUSCLE METABOLISM AND PHYSIOLOGY

Background

Skeletal muscle, which is striated muscle, is composed of bundles of fibers that can shorten or lengthen as necessary. These muscles function in support, locomotion, and work. Additionally, protein in these muscles provides fuel for the body during prolonged food deprivation or following severe injury. During work production, the total muscle force, which reflects the actual power exerted in producing motion or overcoming opposition, depends on the sum of all fibers. Strength, which is defined as the inherent capacity of the muscle, is related to the cross-sectional area of these fibers. Hence, a decrease in fiber size and/or number will affect the work capacity of a muscle. Energy for muscle function is derived largely from glycogen and fats. Glycogen is stored in muscle and is readily available. Therefore, glycogen is used when work begins, followed later by fats from the blood.

Muscles that assist us in standing consume about 15% more energy than when the body is in a supine position. Not all muscles depend on gravity for their function, and the role of these muscles does not change in microgravity. Muscles that depend on gravity are termed "antigravity" and are located generally in the legs, back, face, and neck. These muscles have different types of fibers than nonpostural muscles. In the absence of an antigravity role, muscles may revert to alternate types of fibers.

Spaceflight and Simulation Studies

Measurements from various missions show space travel causes atrophy (wasting) of certain muscles. About 15% of weight loss is due to loss of muscle mass. Muscles whose functions do not depend on bearing weight (e.g., arm muscles) fail to lose mass. Therefore, a large loss by a select group of muscles must account for the bulk of the responses. A recent spaceflight study (STS-48, September 1991) of muscle in young rats showed that some weight-bearing muscles simply grow slower, rather than atrophy. Thus muscle response is complicated by preexisting physiological status.

In humans, the lower extremities lose volume, a third of which is muscle atrophy, but the arms maintain their volume. Indeed, in microgravity, the function of the arms changes to one of locomotion and stability. These results suggest that the loss of muscle mass, especially in the legs, is due to mechanical factors.

Muscle atrophy is associated with breakdown of muscle protein, revealed by the excretion of break-down products, various nitrogen-containing compounds, in the urine. Because nitrogen intake is not increased, the accelerated excretion must result from an excess of the breakdown of muscle proteins over their formation.

Wasting produces physiological changes in muscle. Generally, these adaptations are not a problem until the muscles must function again under the influence of gravity. With extended periods in space, the problems become more severe. Cosmonauts exposed to eight months or more of microgravity generally have difficulty in walking and in maintaining proper posture. Microgravity causes a decrease in muscle tone, work capacity, efficiency, and strength, with an increase in fatigability. Recovery from the effects of space travel varies with the duration of the mission. Muscle strength may require from several days to several weeks to return to normal (i.e., 1 g).

Bedrest studies have been used to test effects of unloading on the musculoskeletal system. Bedrest is not a true simulation because of the ever-present influence of gravity; however, results from such studies are remarkably similar to what we have learned in space. With bedrest, and as is well-known from bedridden patients, there is loss of leg mass and volume, as well as a decrease in the cross-sectional area of muscle fibers and a change in fiber type of the antigravity muscles. Bedrest studies have provided an opportunity to compare the effects of muscle unloading in men and women. Still, such models are limited by being conducted under the influence of 1 g.

Many of the advances in our knowledge of the biochemistry of muscle atrophy have come about by using animal models, as mentioned in the Introduction. Studies on Earth have used hindlimb unweighting to mimic the effects of microgravity on muscle and bone. Limited studies have also been conducted on rats subjected to microgravity. In any event, animals represent a major part of research on the effects of hypokinesia and hypodynamia.

Associated with muscle atrophy are significant physiological changes in the affected muscles, including increased fatigability; decreased strength, elasticity, and force; smaller fiber cross-sectional areas; and change in fiber type of antigravity muscles. Therefore, physiological changes in unloaded rat muscle parallel at least some of those found for unloaded human muscle.

Biochemical studies on unloaded rat muscle have considered the adaptation of carbohydrate, amino acid, and protein metabolism to this intervention. With decreased muscle use, there is a buildup of glycogen, presumably because of decreased utilization of this fuel. The unloaded muscle may respond more to insulin, which protects the muscle to a small extent against loss of protein, accounting for the prolonged period of slower muscle loss following the initial rapid atrophy. Finally, measurements of protein formation and protein breakdown in the unloaded muscle using the animal model suggest that abnormalities in both processes may contribute to the muscle wasting.

BONE AND MINERAL METABOLISM

Background

Bone provides mechanical support to the body and plays an important role in the regulation of body calcium and phosphorus. While it is clear that biomechanical activity controls the interaction of bone and calcium, the mechanisms related to bone growth, mineralization, and maintenance are still poorly understood. Bone mass is in part maintained by loading the bone under the influence of gravity. Muscle tension on bone is also of importance. Bone, like muscle, adapts to limb immobilization and/or unloading. In adult humans, removing the load leads to osteopenia (bone "wasting"), which eventually can cause decreased bone strength and a reduced ability of fractures to heal. The primary bone mineral is calcium phosphate. The bone matrix is primarily protein, of which 95% is collagen. Collagen is unique in that it contains a large proportion of the amino acid hydroxyproline.

Our concepts of the response of the skeletal system to gravitational loading are changing as more information becomes available. During spaceflight, bone responds to an environment where movement of the body and the loads imposed during movement are different from those on Earth. The response to these changes are adaptive, not pathological. The adaptive response to spaceflight involves the entire skeleton, and different parts of the skeleton respond differently. Skeletal adaptation is determined by loading history which, in turn, is a function of exercise, body mass (weight), muscle forces, and fluid pressure and distribution. The ability to adjust to changes in mechanical loading is dependent not only on loading history, but also on normal hormone levels and nutritional intake.

Spaceflight and Simulation Studies

Our current theory suggests that spaceflight, concomitant with the near lack of body weight and the changes in body movements, changes muscle mass. These adjustments, along with fluid shifts, which probably modify blood flow to tissues, cause changes in mechanical loading. The skeleton somehow senses the altered load and adapts to its altered function. The adaptation varies from site to site within the skeleton,

depending on the change in the loading history. The head and possibly the arms may accumulate mineral while the legs and the trunk lose mass. Bones in which maintenance of mineral is gravity-dependent (i.e., due to ground reaction forces) lose mineral, and individuals who are exercisers will probably lose mass in these bones more rapidly than sedentary individuals. Bone loss will occur generally coincident with, though on a slower time scale than, muscle atrophy and decreased muscle strength. These skeletal changes alter the calcium fluxes in and out of bone. The net result of this process is an increase in serum calcium, which initiates a hormone cascade. The increase in serum calcium is usually attributed to increased bone resorption. However, the response is more likely a combination of decreased mineralization with less calcium going into bone and site-specific increases in resorption with increases in calcium fluxes from the bone.

The skeleton adapts to spaceflight as long as diet and endocrine (hormonal) milieu are adequate. The adaptation is a normal physiological process and not a disease state. The result of the adaptation is a change in bone mass with altered architecture and composition. The altered architecture may be reflective of the functional changes of bone and may contribute to the changes in bone strength. However, this adaptation could impair the return to a 1 g environment. The role of bone in mineral homeostasis probably does not change. Spaceflight may be a unique environment to study perturbation of the mineral reservoir independent of loading effects.

The term "bone loss" has been used to describe spaceflight skeletal adaptation and is useful in conveying a site-specific response that triggers the systemic calcium-endocrine response, but the phrase is not appropriate for describing the entire skeleton. We suggest that "bone or skeletal adaptation process" more accurately describes the skeletal response to spaceflight.

Various measurements have been used to estimate bone adaptation. Increases in urine and fecal calcium provide one form of evaluation, though such measurements are affected by the factors described above. Measurements of bone density have been done primarily on the heel bone because of the tremendous load and stress placed on this bone under the influence of gravity. The density diminishes in proportion to the duration of the mission, though a large variability implies that other factors are of importance. For instance, the extent of adaptation may depend on the initial turnover rate of mineral content for each crewmember. Thus, prior physiological condition is an important determinant in the adaptation process.

Much of our understanding of the mechanisms of bone adaptation with unloading have come from bedrest simulations. Such studies showed a loss of minerals similar to that observed during spaceflight, especially in weight-bearing bones. No decrease in density was observed in bones of the upper extremities. While bedrest unloading does not precisely mimic microgravity unloading, simulation studies have allowed prediction of the rate and extent of mineral loss by certain bones.

Animal studies have been conducted in space or using unloading models, such as the couched monkey and the hindlimb-unweighted rat. Monkeys, like humans, show decreased bone mineral and evidence of increased bone resorption. In rats, however, the primary reason for decreased bone mass is a reduced rate of formation without significant resorption. This decreased formation in turn leads to slower growth, demineralization, and decreased bending strength. The different response of bone in rats than in humans or monkeys could be due to a different type of bone structure or to differences in limb motion and loading, or because adult rats show constant bone growth, unlike in adult humans.

Animal models are critical for answering specific questions about the biochemical and physiological adaptation of bone to unloading. Because the rat model does not mimic the response of human bone, it will be important to use other animal models to study the mechanisms of skeletal and calcium changes in space. Still, flight data are critical for validating these models. It is also essential to develop noninvasive analytical methods to facilitate the study of bone loss.

CARDIOVASCULAR AND PULMONARY

Background

The cardiovascular system transports and distributes essential substances (e.g., oxygen and nutrients) to the tissues and removes metabolic by-products. In addition, it contributes to the regulation of body temperature, to hormonal communication within the body, and to the exchange of materials, via the lungs, kidneys, and skin, with the external environment. The cardiovascular system includes the heart, blood, and blood vessels. Blood transports the essential substances described above and is comprised of plasma (the fluid component) and cells. The vessels (vasculature) include arteries, veins, and capillaries. The heart and the vasculature are divided into the systemic (body) and pulmonary (lungs) circulations. The amount of blood the heart pumps (cardiac output), the pressure exerted on the vessel walls (blood pressure), and other aspects of cardiovascular function are exquisitely controlled. Local blood flow is influenced

by a variety of factors including the concentrations of oxygen, carbon dioxide, pH, and metabolic compounds.

Cardiovascular Adaptation to Spaceflight

Under gravitational influence upon standing, blood accumulates in the lower extremities. This pooling increases pressure in vessels below the heart and decreases pressure above the heart. To ensure adequate blood flow to all areas of the body, especially the brain, mechanisms exist to adjust blood flow relative to gravity. Adaptations that occur in microgravity can help us to better understand the normal influences of gravity on Earth. Microgravity alters fluid gradients within the cardiovascular system such that fluid redistributes from the lower extremities to the head, neck, and torso. This redistribution of blood promotes diuresis (fluid loss through urine), which occurs primarily during the first day and continues for up to four days. Diuresis is primarily controlled by a decrease of antidiuretic hormone (ADH). ADH release from the posterior pituitary gland is regulated by a nerve signal from the right atria of the heart. To maintain normal plasma osmotic pressure, sodium (electrolyte) loss must be commensurate with fluid loss.

Plasma electrolytes are regulated by the renin-angiotensin-aldosterone hormone cascade. Weightlessness decreases fluid pressure and nerve signals to the kidney, thus reducing renin release. Consequently, angiotensin formation and aldosterone release are decreased due to less renin. Aldosterone enhances sodium reabsorption in the kidney, such that aldosterone decrease can lead to a marked sodium loss.

Despite the limited inflight experimentation, it is clear that the cardiovascular system undergoes marked physiological alterations in response to the fluid shifts experienced in weightlessness. Changes in muscle tone of the vasculature maintain adequate blood flow in spite of the reductions in blood volume during weightlessness. A general cardiac deconditioning, relative to 1 g status, occurs in flight. Consequently, there are marked alterations in various physiological parameters, such as increased heart rate and mean arterial blood pressure, during exercise.

In addition to functional changes in the heart and vessels of the cardiovascular system, blood components are altered by spaceflight. For instance, red blood cells (RBC) are decreased during spaceflight. This reduction results from either decreased production or increased destruction of the cells. Serum erythropoietin, which stimulates RBC production, is decreased during flight. The increase of serum ferritin in weightlessness indicates increased spleen breakdown of RBC. The reduction in plasma volume may also contribute to reduced RBC mass. Since RBC are essential for carrying oxygen, this has important implications in getting oxygen to tissues upon return to normal physiological conditions.

A cardiovascular response that may not be directly linked to fluid shifts during spaceflight is increased incidence of dysrhythmias (abnormal heart beats) observed in many crewmembers. While no definitive cause for spaceflight dysrhythmias has been established, probable factors include gravitational stress (e.g., during reentry), thermal loads (e.g., during extravehicular activity), and electrolyte or hormone alterations.

Significant cardiovascular changes occur as a physiological adaptation to a novel environment. They are not inherently deleterious in microgravity but create the problem of deconditioning upon return to gravitational influence. One of the most significant problems is orthostatic intolerance, the inability to maintain adequate cardiovascular function while standing under the influence of gravity. Postflight tests show increased heart rate and decreased pulse pressure as compared with preflight measurements. Additionally, increased leg volume during orthostatic tests were greater postflight than preflight, suggesting inability to regulate vascular tone following weightlessness. Decrease in exercise capacity is also manifest after spaceflight.

Pulmonary Adaptation to Microgravity

The pulmonary system exchanges gases with the blood and contributes to the regulation of acid-base balance, which is critical for survival — a sharp shift towards acid or base can be fatal. Anatomically, the respiratory system is comprised of the lungs and the pulmonary circulation. Under the influence of gravity, gradients are established in the lung for gas volume (ventilation) and blood flow (perfusion). These gradients result in greater ventilation and perfusion in the bottom of the lung. The ratio of ventilation to perfusion determines the amount of gas exchange between the air and blood in a given portion of the lung. Therefore, the exchange of gas is best in the lower portions of the lung.

Because of limited data, it is difficult to ascertain the influence of microgravity on the pulmonary system. There have been no reports of postflight abnormalities. Hypothetically, microgravity should alter lung distention, ventilation, and ventilation-perfusion ratios, thus improving gas exchange at rest. Chronic changes in ventilation-perfusion ratios may also affect the function of the right heart. Microgravity effects on maximal oxygen consumption are of the

most vital concern because of its potential to limit work capacity in space and upon readaptation to 1 g.

NEUROVESTIBULAR

Background

The neurovestibular system controls spatial orientation, coordinated motor performance, and postural maintenance with respect to gravity. Information from specialized organs in the inner ear, along with input from sensory pathways, is integrated in the central nervous system, the brain, to complete these tasks.

The primary structures for obtaining information regarding linear acceleration and the direction of the gravity vector are the otolith organs. These organs contain hair cells embedded in a gelatinous mass containing calcium carbonate crystals called otoconia. Changes in head orientation or linear acceleration impart forces on the otoconia resulting in altered electrical discharge from the hair cells. In addition, the hair cells provide background electrical discharge commensurate with the force of gravity exerted on the otoconia.

The detection of angular acceleration is accomplished by three semicircular canals, accounting for the three planes of orientation. Angular acceleration of the head results in fluid streaming in the semicircular canals corresponding to the plane of movement, ultimately producing an electrical discharge in cells of that canal. Signals from the various sensory organs contribute to the pool of information that is integrated in the brain. This information produces a coordinated signal from the brain to the skeletal and eye muscles. Additionally, there is output to areas of the brain responsible for controlling digestion, blood pressure, and respiration. Pressure sensations induced by gravity alter tactile responses during flight, which may also contribute to postural and spatial orientation alterations.

Adaptation to Spaceflight

In microgravity, the neurovestibular system must adapt to an altered set of sensory cues, which result in acute changes in the output from, or integration within, the neurovestibular system. Specific adaptations, considered below, include changes in spatial orientation, postural maintenance, the vestibulo-ocular reflex (VOR), and central processing in the neurovestibular system.

Spatial orientation describes the relationship between the body and an external reference frame and is accomplished by comparing a variety of external inputs. Spaceflight can lead to impaired integration due to the lack of gravitational effects on the otolith organs. This may result in a sudden reversal of orientation, the so-called inversion illusion. While the semicircular canals are relatively unaffected by microgravity, the detection of static head positions by the otolith organs may be impaired. This alteration may lead to spatial disorientation, resulting in an increased dependence on static visual cues, the use of tactile cues (e.g., from the soles of the feet) to yield an upright sensation, and the alignment of the perceived vertical axis with the long axis of the body.

Maintenance of posture and equilibrium requires integration of information from visual, vestibular, and somatosensory systems. This information coordinates muscular activity to orient the body with respect to gravity. Under gravitational influence, set patterns of muscle activity are the strategy used to adjust automatically the center of gravity to a stable position following perturbations to the body. In microgravity, the sensory interpretation and muscular coordination are changed. These changes are thought to represent altered strategies of response such that the patterns of muscle activity established under gravitational force are changed.

The VOR provides stable vision during head movement. Ocular compensation to head motion is accomplished through a pathway between the semicircular canals and the muscles of the eye. Currently, little is known about the adaptation of the VOR to weightlessness. It is hypothesized that the disparity between sensory input from various sources may result in acute disorientation and motion sickness.

While the adaptations to microgravity render the neurovestibular system well suited for the environment of weightlessness, returning to gravitational influence requires significant readaptation. At present, there is little information describing this process. Alterations in the detection of linear acceleration, a continued increase in the dependence on visual cues and illusions, such as floor motion during vertical movement, occur during the period following weightlessness prior to readaptation. The degree and duration of such symptoms are probably proportional to mission length.

LIVER

The liver is the "manufacturing plant" of the body producing glucose, blood proteins, and lipids as needed. Additionally, it is the site at which the body removes drugs from the circulatory system. Yet the potential response of the liver to spaceflight is very unclear as there is no useful Earth-bound model that can mimic potential effects on this organ. Some limited data have been obtained from rats flown on Spacelab 3. Undoubtedly, future studies are essential to ascertain whether spaceflight affects the function of this critical organ.

ENDOCRINE ORGANS

Background

Many of an organism's adaptations to change in physiological status are related to responses of those that release hormones as part of the endocrine system. These may include the adrenal, pituitary, and thyroid glands, and the pancreas. Hormone balance in the blood controls body metabolism. Response of part of the endocrine system was discussed above in conjunction with the renin-angiotensin-aldosterone response and the change in antidiuretic hormone. Other important hormones include insulin, which promotes fuel storage and maintenance of body protein; cortisol (a glucocorticoid), which promotes the release of fuel from storage; thyroid hormone, which regulates body metabolism by increasing energy production; and growth hormone, which promotes tissue growth and repair. Imbalance of these hormones with altered physiological status has serious implications when the body attempts to respond to stress insults such as injury.

Spaceflight and Simulation Studies

An increase of blood cortisol occurs in response to stress. Both spaceflight and bedrest simulation increase the amount of cortisol. Another indicator of stress is increased human growth hormone. However, release of growth hormone from the pituitary may be suppressed during spaceflight, as suggested by a study using rat pituitary glands and cells. Indeed, spaceflight may produce some direct effect on growth hormone producing cells in this endocrine gland.

Insulin, which normally counteracts the glucocorticoid (cortisol) action, was found to be diminished after two weeks of spaceflight. Possible loss of this antagonistic effect of insulin could have significant implications for the maintenance of tissue and organ size and the ability to respond to injury in space.

Thyroid hormone, which increases oxygen consumption and heat production, may be increased by spaceflight. This finding is in keeping with elevated oxygen consumption and the high energy demands in space. Coupled with decreased red blood cell mass for transporting oxygen to tissue and organs, there could be serious ramifications of these opposing responses.

We have just begun identifying the consequences of space travel on hormone balance. A failure to follow these studies to completion could seriously impair our ability to maintain humans in space for long periods of time. Inflight measurements on humans and animals will be essential because postflight sampling is likely affected by readaptation to gravitational force. Use of animal and human Earth-bound model systems also seems fruitless for these studies as it is not yet possible to mimic the specific hormonal pattern of spaceflight.

BASIC SCIENCE QUESTIONS

1. What role does gravity play in the development of support structures such as bone and muscle?

2. What is the influence of gravity and its lack on the formation, turnover, and metabolism of support structures?

3. What is the relative importance of altered load bearing and gravitational force in metabolic adaptations of muscle and bone?

4. What is the role of the endocrine system in the response of the support structures to gravitational influence?

5. What role does gravity play in development of bone strength and muscle physiology or the lack of gravity in diminution of these physiological parameters?

6. Is it possible to find a mechanical or electrical perturbation that can substitute for gravity for the development of support structures? If so, are the responses to artificial gravity equivalent to that on Earth?

7. How does gravity influence biomineralization?

8. How does gravity influence physiological systems such as cardiovascular, pulmonary, endocrine, and neurovestibular?

9. To what extent will alteration of organs and tissues in space lead to impaired response? Will any response be pathological, or are they simply adaptive?

10. To what extent are alterations in microgravity reversible upon return to 1 g?

RESEARCH PRIORITIES

1. Determine the validity of the partially unloaded rat and human bedrest models for predicting spaceflight changes, both short-term and long-term duration, in organs and tissues, especially bone and muscle, and for understanding the basic mechanisms of these changes.

2. Study the dynamic role of calcium in gravity-mediated responses of bone.

3. Determine how muscle tension or mechanical strain influence bone growth during skeletal unloading, and whether altered bone growth affects the extent of muscle growth. Such studies should consider interrelation of movement, muscle tension, posture, and skeletal strength.

4. Use the microgravity environment to understand how organisms have adapted their structure to withstand the gravitational influence of Earth during evolution.

5. Determine whether bone crystal size, form, or defect sites are altered by unloading.

6. Determine the mechanism for muscle atrophy and/or altered growth with unloading.

7. Identify the precise pattern of endocrine changes with spaceflight and the ramifications of these changes in terms of organ and tissue functions.

8. Use the microgravity environment to understand how organisms adapted their control (regulation) of organ/tissue function during evolution and how they adapt to changes in their gravitational environment.

9. Dissect the possible components of the gravitational influence to evaluate which is the major contributing factor in each adaptive response.

BIBLIOGRAPHY

Committee on Space Biology and Medicine. 1987. *A Strategy for Space Biology and Medical Science for the 1980s and 1990s* (The Goldberg Report). Washington, DC: National Academy Press, 220 p.

Desplanches, D., Mayet, M.H., Ilyina-Kakueva, E.I., Sempore, B., and Flandrois, R. 1990. Skeletal muscle adaptation in rats flown in Cosmos 1667. *Journal of Applied Physiology* 68(1): 48-52.

Grindeland, R., Hymer, W.C., Farrington, M., Fast, T., Hayes, C., Motter, K., Patil, L., and Vasques, M. 1987. Changes in pituitary growth hormone cells prepared from rats flown on Spacelab 3. *American Journal of Physiology* 252(2, Part 2): R209-R215.

Grindeland, R.E., Ed. 1990. Cosmos 1887. *FASEB Journal* 4(1): 10-109.

Holy, X. and Mounier, Y. 1991. Effects of short spaceflights on mechanical characteristics of rat muscles. *Muscle & Nerve* 14(1): 70-78.

Martin, T.P. 1988. Protein and collagen content of rat skeletal muscle following space flight. *Cell and Tissue Research* 254(1): 251-253.

Merrill, A.H., Wang, E., Jones, D.P., and Hargrove, J.L. 1987. Hepatic function in rats after spaceflight: effects on lipids, glycogen, and enzymes. *American Journal of Physiology* 252(2, Part 2): R222-R226.

Morey, E.R. and Baylink, D.J. 1978. Inhibition of bone formation during space flight. *Science* 201(4361): 1138-1141.

Morey-Holton, E. and Tischler, M. (Eds.) 1988. *NASA Workshop on Biological Adaptation*. Moffett Field, CA: NASA, Ames Research Center, 102 p. (NASA TM-89468)

Morey-Holton, E.R. and Arnaud, S.B. 1991. Skeletal responses to spaceflight. In: *Advances in Space Biology and Medicine*, Vol. 1. (Bonting, S.L., Ed.). Greenwich, CT: JAI Press, p. 37-69.

Musacchia, X.J. and Steffen, J.M. 1984. Cardiovascular and hormonal (aldosterone) responses in a rat model which mimics response to weightlessness. *Physiologist* 27(6): S41-S42.

Musacchia, X.J. and Steffen, J.M. 1982. Short term (1 and 3 day) cardiovascular adjustments to suspension antiorthostasis in rats. *Physiologist* 25(6): S163-S164.

Musacchia, X.J., Steffen, J.M., Fell, R.D., and Dombrowski, M.J. 1990. Skeletal muscle response to spaceflight, suspension, and recovery in rats. *Journal of Applied Physiology* 69(6): 2248-2253.

Nicogossian, A.E., Huntoon, C.L., and Pool, S.L. (Eds.) 1989. *Space Physiology and Medicine*, 2nd Edition. Philadelphia: Lea & Febiger, 421 p.

Oganov, V.S., Rakhmanov, A.S., Novikov, V.E., et al. 1991. The state of human bone tissue during space flight. *Acta Astronautica* 23: 129-133.

Riley, D.A., Ellis, S., Slocum, G.R., Satyanarayana, T., Bain, J.L.W., and Sedlak, F.R. 1987. Hypogravity-induced atrophy of rat soleus and extensor digitorum longus muscles. *Muscle & Nerve* 10: 560-568.

Sandler, H. and Vernikos, J. (Eds.) 1986. *Inactivity: Physiological Effects*. Orlando, FL: Academic Press, 205 p.

Steffen, J.M. and Musacchia, X.J. 1986. Spaceflight effects on adult rat muscle protein, nucleic acids, and amino acids. *American Journal of Physiology* 251: R1059-R1063.

Task Group on Life Sciences. 1988. *Space Science in the Twenty-First Century: Imperatives for 1995-2015 Life Sciences*. Washington, DC: National Academy Press.

Tischler, M.E., Jaspers, S.R., Henriksen, E.J., and Jacobs, S. 1985. Responses of skeletal muscle to unloading — A review. *Physiologist* 28(6): S13-S16.

Wronski, T.J. and Morey, E.R. 1983. Alterations in calcium homeostasis and bone during actual and simulated space flight. *Medicine and Science in Sports and Exercise* 15(5): 410-414.

Wronski, T.J and Morey, E.R. 1983. Effect of spaceflight on periosteal bone formation in rats. *American Journal of Physiology* 244(3): R305-R309.

A Scientific Role for Space Station Freedom: Research at the Cellular Level

TERRY C. JOHNSON
Division of Biology, Kansas State University, Manhattan, KS 66506

JOHN N. BRADY
Division of Cancer Etiology, National Cancer Institute, Bethesda, MD 20892

ABSTRACT

The scientific importance of Space Station Freedom is discussed in light of the valuable information that can be gained in cellular and developmental biology with regard to the microgravity environment on the cellular cytoskeleton, cellular responses to extracellular signal molecules, morphology, events associated with cell division, and cellular physiology. Examples of studies in basic cell biology, as well as their potential importance to concerns for future enabling strategies, are presented.

INTRODUCTION

We are at the threshold of a historic opportunity to explore the potential role of gravity and the biological responses, at a cellular level, to the microgravity environment. Since the activities and properties of all organs and tissues, of both plants and animals, are communal expressions of their cell components, cell biology lies at the basis of all life forms.

While gravitational forces can be experimentally increased and almost every other aspect of the life environment of plant and animal species controlled, the potential impact of Earth's gravity on living cells, tissues, and organ systems remains largely unknown.

There at least four major reasons for studying gravitational biology at the cellular level: (1) to gain fundamental knowledge of the potential influences of the microgravity environment on the cellular functions of both plant and animal cells; (2) to relate the cellular activities, altered under gravity unloading conditions, to a better understanding of events on Earth — in unit gravity — that are associated with the regulation of cell proliferation, gene action, development, etc.; (3) to exploit altered functions that occur in microgravity to generate products that will improve the quality of life; and, (4) to provide accurate projections of those long-term influences of the microgravity environment on cellular functions that may threaten future space exploration (Life Sciences Division Working Group, 1991).

Although gravitational cell biology is in its infancy, there are clearly numerous guideposts that indicate that the future holds many interesting surprises, both pleasant and unpleasant, regarding how both plant and animal cells will respond to gravitational unloading, whether space adaptation is possible at the cellular level, and what physiological processes in the intact species will be significantly altered as a result of cellular responses to reduced gravity.

The major barrier that is presently faced by gravitational biologists is the scarcity of flight opportunities available for scientific research. Of no less importance is the relatively brief duration that characterizes our opportunities for microgravity research. Brief parabolic episodes on aircraft offer valuable, but extremely limited, opportunities for biological research. In many cases, these flights offer little more than opportunities to test various flight hardware and to test concepts of experimental design. Orbiter flights have had durations of only a few days and minimal opportunities, with but a brief time available by busy crewmembers with their manifold responsibilities in flight to aid in experimentation, and frequent launch delays simply constrain many experimental designs with living cells.

Gravitational biology may mature as a science only when a dedicated science laboratory, like Space Station Freedom, is available for intensive and long-duration studies. A manned space station can be neither justified nor denied for reasons other than the scientific potential it promises. In fact, no scientific facility is necessary as an end in itself. It is the applications to future scientific advancement that drive the need for any new instrumentation in science and engineering. For instance, is there a need for the electron microscope to allow advancement of science? The answer is, of course, both yes and no. Scientific advancements can, and are, made in many areas of biological sciences without the use of the electron microscope. However, if scientific progress in other areas requires a visual examination of cellular substructure, beyond the resolution of the light microscope, the electron microscope is an indispensable instrument. In a similar vein, there are several ways to study gravity unloading with cellular systems. However, there is no question that progress will continue to be slow, incomplete, and excruciating in the absence of a permanently orbiting scientific laboratory.

It is reasonable to expect the biological scientific community to provide justification for the significant international investment that will be required for the development and maintenance of Space Station Freedom. Perhaps the most convincing evidence can be found in the intriguing observations of cellular responses to gravity unloading that have already been made under less than satisfactory experimental opportunities.

It is already clear that microgravity has an impact on living systems, including normal gravitropic responses of plant root tips, the developmental program of amphibian species, the embryological development of certain avians, and the growth dynamics of unicellular microbes. We suspect, however, that the influence of unit gravity is much more pervasive than presently known, and that the absence of Earth's gravitational forces will have a magnitude of influence that may be a barrier to long-term survival of life forms, including human, in the hostile space environment.

In some cases cells may serve as gravity sensors, although in most cases cells are influenced primarily by the microgravity environment and the absence or reduction of buoyancy-driven microconvection currents. Many of these features have already been extensively reviewed (Halstead and Dutcher, 1987; Todd, 1989; Krikorian and Levine, 1991; Space Studies Board, 1991; Lewis and Hughes-Fulford, 1993), and the present report will focus on newer observations, concentrate on cellular activities that are more indirectly influenced by the microgravity environment, and provide a synthesis of the potential impact that this influence may have on the future of space life sciences.

GRAVITATIONAL BIOLOGY AT THE CELLULAR LEVEL

The overall objectives of gravitational biology at the cellular level encompass identification of cellular processes uniquely influenced by the full spectrum of gravitational forces, and the access by researchers to g forces equal to and less than unit gravity (Life Sciences Division Working Group, 1991).

The goals include measures to:

(1) Identify how single cells sense gravity, including both direct and indirect (environmentally mediated) effects.

(2) Identify how cells transduce gravitational stimuli and how they respond to both acute and long-term variations in gravitational force.

(3) Develop model cell systems to describe processes and mechanisms by which cells respond to altered gravitational force.

THE EXTRACELLULAR MATRIX AND CYTOSKELETON AS A GRAVITATIONAL REACTIVE COMPLEX

Others have suggested that the presence or absence of gravitational forces may influence cell function by modifying structural components associated with the extracellular matrix, plasma membrane, and cytoskeleton (Ex-Me-Cy) (Todd, 1989; Cipriano, 1990; Spooner, 1992) (Figure 1). Table I lists some of the components of the Ex-Me-Cy. These cellular substructures represent large polymerized complexes and each is characterized by a considerable macromolecular instability and, therefore, continuous turnover of their macromolecular components. Their significant size, intermolecular interactions, and turnover rates provide properties that suggest that they are reasonable candidates for being influenced by gravitational forces (Todd, 1989). Furthermore, they are involved in a wide range of cellular and intercellular features, including: cell-cell communication; cellular attachment and aggregation; cellular morphology; signal transduction; cellular contractile properties and motility; endocytosis; exocytosis; ion fluxes; and, molecular interactions with the proteins, glycoproteins, and lipids that comprise the architecture of the fluid mosaic membrane.

The extracellular matrix and the cytoskeleton are integral complexes common to both plant and animal cells. Although the extracellular matrix and cytoskeleton are often considered independent entities, they are in fact intimately associated directly and indirectly through their organization with both surface and intracellular membranes. The extracellular matrix of plant cells is distinguished by celluloses, pectins, and lignins comprising the cell wall. The extracellular matrix of animals cells is distinguished by collagens, proteoglycans, and laminins. The cytoplasmic cytoskeleton of all cells is established by an array of interconnecting microtubules, intermediate filaments, and microfilaments. These structures are

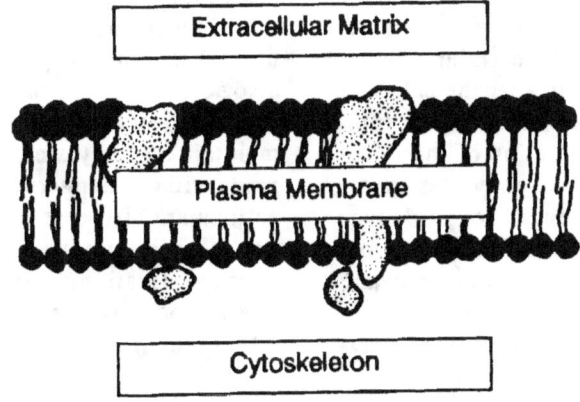

Figure 1. The Extracellular Matrix - Plasma Membrane - Cytoskeleton Complex.

Table I. Examples of the molecular components of the Ex-Me-Cy.

Extracellular Material	Plasma Membrane	Cytoskeleton
Collagens	Proteins	Microtubules
Proteoglycans	Glycoproteins	Microfilaments
Laminin	Lipids	Intermediate filaments
Fibronectin	Phospholipids	
Celluloses		
Pectins		
Lignins		

separated by a plasma membrane that provides a wide arrangement of binding sites for interactions with both the extracellular matrix and cytoskeletal complexes. In reality, therefore, these large polymerized substructures act in concert, as an integrated lattice, to regulate and influence a remarkable range of cellular activities. Many of these cellular activities are fundamental to the survival and function of the plant or animal, particularly in plant and animal tissues where the differentiated cellular components share a common overall purpose, yet bear distinctly separate responsibilities.

Many, if not most, of the cellular perturbations associated with the microgravity environment, therefore, may be consequences that will ultimately be associated with altered extracellular Ex-Me-Cy interactions. A brief overview of critical consequences associated with a prolonged exposure of cells to the microgravity environment includes events affiliated with cell-cell interactions and signal transduction, cell division, and immune responses.

Although most likely an indirect physiological consequence to gravity unloading, a microgravity response that involved the elements of the Ex-Me-Cy ultrastructure has been reported in animals subjected to a microgravity environment. Experimental observations from cellular elements of animal tissues from the Spacelab 3 mission in 1985 described cellular changes in tubulin and cytoskeleton synthesis and distribution, and changes in collagen secretion (Space Studies Board, 1991). Ultrastructural data from rats exposed to microgravity for 12.5 days on the Cosmos 1887 flight illustrated that perturbations in protofibrils (actin and myosin filaments) of rat cardiac tissue could result from gravity unloading (Philpott et al., 1990). Related observations were made on marked reduced myofibril yields from vastus intermedius muscles of rats from Cosmos 1887 (Baldwin et al., 1990). Interestingly, cytoskeletal elements associated with neuromuscular junctions have been shown to be altered when neurons and myocytes were cultured in a vector-free gravity environment (Gruener, 1991).

KC-135 and Consort I sounding rocket flights have been used to determine if a reduced gravity environment can influence assemblies of macromolecules that are associated with the cellular Ex-Me-Cy complex. Reduced gravity did alter the cell-free assembly of tubulin, collagen, and fibrin clot formation although the degree and direction of the influence was different for each of the molecules examined (Moos et al., 1990). Although these cell-free results may have been primarily a reflection of flow dynamics in the reduced gravity environment, and may not have been directly applicable to the intracellular cytoskeleton, flow dynamics in reduced gravity may play a key role in the maintenance and turnover of the extracellular matrix. In turn, these extracellular matrix changes could alter the molecular relationship of the cytoskeleton to the plasma membrane. It is also conceivable that the subsequent changes in the plasma membrane could transduce alterations in the cytoskeleton as it is associated with the cell surface.

These and other observations regarding the potential influence of a reduced gravity environment on the Ex-Me-Cy complex harbor potential consequences for an unusual number of cellular activities that are essential for maintaining biological integrity. In addition to the interesting basic science questions concerning gravitational impacts on living biological species, one cannot help but harbor concern that serious alterations in these functions may compromise the long-term survival of many biological species in the microgravity environment.

EXAMPLES OF CELLULAR ACTIVITIES THAT COULD BE COMPROMISED BY GRAVITATIONAL INFLUENCES ON THE EX-ME-CY COMPLEX

The cell surface plays a key role in cellular responses to extracellular cues such as peptide hormones, growth factors, growth inhibitors, and a myriad of signal molecules that provide critical information to this sensing organelle. In addition, direct cell-cell communication is mediated by macromolecules on the cell surface, and the Ex-Me-Cy complex responds to these signals in a manner that mediates cellular decisions that are essential to survival.

Cell-Cell Interactions

Cell-cell interactions play key roles in cellular communication that range from those that involve direct cell-cell contact to those that involve soluble ligands. Direct cell-cell contact plays an essential role in the formation of cellular aggregation assemblies that are essential for normal tissue and organ development as well as for certain aspects of the normal immune response. Direct cell-cell contact also seems to be an important mechanism in inhibiting cell proliferation and the ability of tissues to maintain a steady-state turnover of cellular components without dangers of hyperplasia. Direct interactions are also essential to plant-microbe interactions that initiate the process that leads to symbiotic nitrogen fixation, and initial studies of this key interaction have been carried out in reduced gravity (DeBell et al., 1990; Urban, 1991).

Soluble ligands are the main communication links between cells over a distance, and these molecular cues provide a wide range of cellular responses, including alterations in cellular metabolism, stimulation of cell division, promotion or discouragement of neoplastic growth, cellular mobilization associated with inflammation, normal wound healing, and many key responses essential to the immune response.

Many features of cell-cell communication involve specific receptors associated with the plasma membrane. However, many of the responses described above also involve alterations in the entire Ex-Me-Cy complex, and are not limited to a simple molecular interaction of a ligand and receptor. For instance, the receptor-ligand complex can initiate a signal transduction cascade that involves the entire Ex-Me-Cy complex. The signal molecules can be components of the extracellular matrix that can interact with linking elements called integrins (Spooner, 1992). The subsequent cellular response may include the synthesis and secretion of extracellular matrix molecules that necessitates the participation of both the cytoskeleton and the plasma membrane.

The fluid mosaic plasma membrane is a dynamic structure that involves movement of its lipid and many of its macromolecular components at a remarkable rate. As reviewed by Edidin (1987), both lipid and protein components have a rotational diffusion that can be measured experimentally, and even single unaggregated 30 to 100 kDa proteins rotate on a time scale of microseconds. Addition of cytoskeletal proteins can increase the rotational correlation time as much as two-fold. Lateral diffusion constants for a wide variety of vertebrate membrane proteins have been measured, and generally are in the range of $D = 5 \times 10^{-9} \, cm^2 \, sec^{-1}$. These values are generally what would be predicted from the viscosity of synthetic phospholipid bilayers or estimated with the rotational diffusion of membrane proteins. However, many measurements of lateral diffusion made with native membrane preparations have been an order of magnitude less than what would be

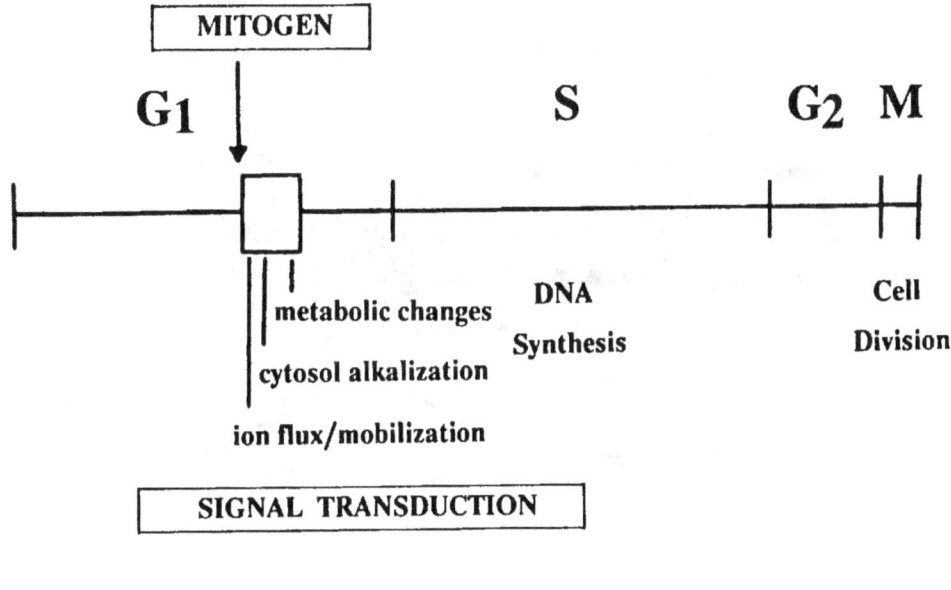

Figure 2. A linear display of the cell cycle and the most common arrest site at G_0/G_1.

predicted, and it is thought that interactions of the plasma membrane with cytoskeleton elements result in this discrepancy. Consistent with this possibility have been measurements of the lateral diffusion dynamics of membrane proteins in native membranes, in the absence of cytoskeletal proteins, that turn out to be 10 to 1000 times faster than that measured in native membranes in the presence of macromolecules of the cytoskeleton (Edidin, 1987). Even molecules diffusing as slowly as 10^{-13} cm^2 sec^{-1} can transverse the cell within a few hours. Of course, all membrane proteins and glycoproteins do not have the same freedom of motion. Many are anchored or clustered by elements of the extracellular matrix and cytoskeleton, and remain relatively quiescent compared to those macromolecules free to diffuse in the inner and outer sides of the lipid bilayer.

Cell Division

There is little question that the decision of a eukaryotic cell to divide or not to divide is a summation of external signals that involves the action of both growth factors and growth inhibitors. Both promoters and inhibitors of cell division influence cellular metabolism by binding to specific cell surface receptors that are most likely residents of the plasma membrane. Growth inhibitory molecules maintain the cells ar-

rested in the G_0/G_1 phase of the cell cycle (Figure 2) and growth factors, when overcoming these inhibitory influences, drive the cells into the S phase where DNA and histones are synthesized in preparation for cell division (Pardee, 1989).

The interaction between inhibitory and stimulatory ligands is extremely complex and remains the subject of intensive ground-based research. There are several pathways involved in signal transduction, as related to the control of cell proliferation, and this will be a key focus for future studies in the reduced gravity environment. In general, the binding of growth factors to the cell surface initiates a metabolic cascade that includes ion fluxes, release of Ca^{2+} from internal membrane stores, an alkalinization of cytosol, metabolism of polyphosphate inositol, and the phosphorylation of cytosolic and nuclear proteins (Figure 3). The cascade includes the induction of specific gene expression that provides a state of competence for subsequent reentry of the cell to the cell cycle (Johnson and Sharifi, 1989; Toole-Simms et al., 1991; Fattaey et al., 1991; Edson et al., 1991).

As previously reviewed (Lewis and Hughes-Fulford, 1991), cell division of both plants and animals does appear to be significantly influenced in the microgravity environment. Several experiments have shown that prokaryotic bacteria and single plant cells appear to proliferate more rapidly in space than ground-level controls. Although there may be a multitude of

25

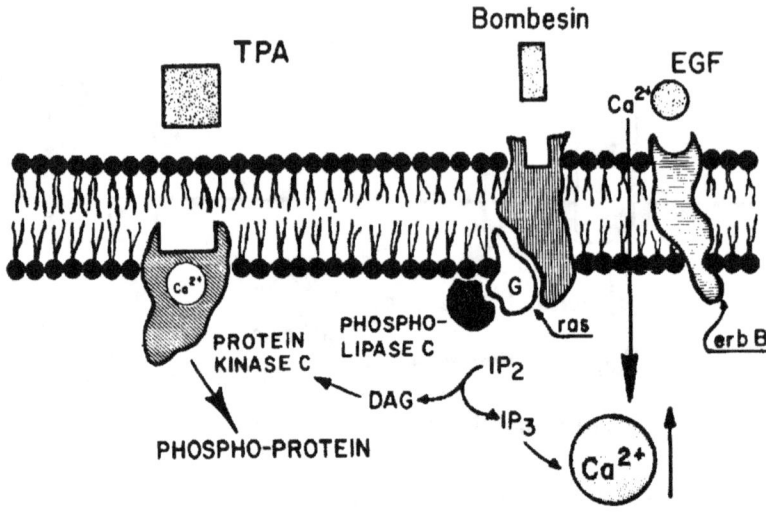

Figure 3. Examples of signal transduction events associated with cell cycling.

complications in the few space-related experiments that have been performed to date, at least some species of higher plants appear to manifest a significantly greater degree of development and differentiation under weightless conditions. Furthermore, the more overt differences between nominal gravity and weightless influences on plant development are manifested at the subcellular level (Lewis and Hughes-Fulford, 1993).

Proliferation of animals cell cultures in microgravity also seems to be different than of ground-based controls although, in general, animal cell proliferation seems to be reduced in the weightless environment. Whether this is a result of altered flow dynamics, mitogen-receptor interactions, events associated with signal transduction, DNA synthesis and chromosome replication, or cytokinesis remains to be determined. Nevertheless, the preliminary observations are both intriguing and hold potential significance to the support of life in the space environment.

Clinostat and sounding rocket experiments have suggested that altered gravity potentially influences the clustering of surface membrane epidermal growth factor (EGF) receptors and subsequent expression of c-fos, a gene associated with the entry of cells into the mitotic cycle (Rijken et al., 1990). The mechanism for such a result is unknown although it is interesting that EGF receptor clustering may be a key facet of mitogenesis by this growth factor. Since both chromosome segregation and cytokinesis involve the participation of microtubules, alterations in tubulin assembly in the microgravity environment would be

expected to have dramatic effects on the latter stages of cell division.

One problem that continues to plague space life scientists is the inability to discriminate between true microgravity influences and other, indirect factors that accompany many space missions. In only very few experimental situations have inflight centrifuges accompanied these experiments to produce unit gravity. For instance, one can reasonably question if orbiter and satellite flight experiments are necessarily providing results about reduced gravity or increased radiation. As flight durations increase, the difficulties in discriminating between these two potential influences, and many others, will only become more complex and difficult. Space Station Freedom would be a unique and imperative laboratory for these future experiments since both unit gravity and weightless cells and tissues could be studied simultaneously.

Immune Cell Activities

Of no less importance is the potential influence of reduced gravity on the immune response. In some respects immune cells offer a compelling model system for studies of cell-cell communication. The overall immune response is characterized by a significant degree of cellular interaction, including both suppressor and stimulatory actions, as well as by responses to both direct cell-cell interactions and soluble ligands (lymphokines) that modulate the overall immune response. Furthermore, the concern that has emerged about the potential of a compromised immune system in reduced gravity, and the myriad of health-related issues that could ensue, makes this an important area of study for the space life sciences.

This area of research has received a great deal of attention and many of the observations have been amply reviewed (Cogoli and Tschopp, 1985; Lewis and Hughes-Fulford, 1993; Sonnenfeld, 1989; Cogoli et al., 1990). Again, the conclusions regarding microgravity and immune cell function are often conflicting and the differences among experimental systems and approaches makes generalization difficult. Nevertheless, there is ample evidence that suggests gravity unloading results in immunologic changes in humans and other vertebrates. In addition to *in vivo* experiments, studies with immune cell suspensions have clearly indicated that the response of the immune system to microgravity may be significant indeed. As discussed by Sonnenfeld (1989), weightlessness, stress, and low-level radiation could all contribute to alterations in the immune response, and what factors are responsible for the measured differences remain to be established.

On one hand, discriminating among the various factors that could influence the immune system remains an important goal for space life sciences in order to more fully describe the basic mechanisms that underlie these altered responses. On the other hand, since a compromised immune system could be a disabling barrier to the long-term health and welfare of humans and other vertebrates in the space environment, understanding changes in immune responses has very practical implications.

Space environment-mediated alterations of the immune system have been known for several years although, until recently, the cellular mechanisms for immune cell changes in reduced gravity have been only sparsely studied. Before the orbiter flights, studies of lymphocytes obtained from astronauts and cosmonauts showed that the mitogen-mediated activation could remain suppressed for days after their return to Earth (Taylor et al., 1986; Gould et al., 1987). However, it has been difficult to discriminate between the possibility that physiological stress-induced changes, rather than a more specific immunological response, was responsible for this immune suppression. A series of benchmark studies, however, illustrated that even lymphocytes in culture could be influenced by reduced gravity, and reduced gravity-mediated changes in the extracellular matrix and/or the plasma membrane have been suggested as a mechanism for this effect (Cogoli et al., 1990).

Although observations with cellular elements of the immune culture would not be expected to necessarily parallel immunological events in the intact host, both *in vivo* and cell culture studies will be necessary to provide a comprehensive understanding of reduced gravity and the immune cell response. A series of studies on cytokines, at the cellular level, has led to numerous interesting observations. Cytokines are molecular messengers that mediate immune cell-cell communication and orchestrate the overall immune response. In a study after a one-week orbiter flight and upon return of experimental rats to Earth, isolated splenocytes were examined for the relative induction of two lymphokines, interferon-gamma and interleukin-3. While the synthesis and release of interferon-gamma was significantly depressed in comparison to ground-based controls, the production of interleukin-3 was not influenced by the reduced gravity environment (Gould et al., 1987). A recent immune cell study, carried out on a KC-135 parabolic flight, used isolated peritoneal macrophages to show that superoxide (O_2^-) production was four-fold higher in reduced gravity when compared to unit gravity control cultures (Fleming et al., 1991). The altered activity of these inflammatory cells could hold a great deal of significance since superoxide is a major component in macrophage-directed bacterial killing. Although the exact mechanism for the higher production of superoxide in reduced gravity is unknown, it has been speculated that the altered cellular component might well be the Ex-Me-Cy complex (Fleming et al., 1991).

In experiments recently conducted with bone marrow-derived macrophage suspensions during orbiter flight and within 12 hours of reduced gravity, over a three-fold increase in interleukin-1 production was measured in comparison to that measured with ground-based macrophage control cultures (Chapes et al., 1991). In addition, the production of tumor necrosis factor (TNF-alpha), another substance important to inflammatory cell activity, was shown to be significantly stimulated by the space environment (Chapes et al., 1991).

Clearly, the immune system and immune cells themselves are influenced by the microgravity environment, although the degree and direction of the response, and the specific lymphokines involved in the response, appear to vary. Undoubtedly, many new observations will emerge from studies of immune cell function in microgravity, and the information gained may prove to be of significant importance to both basic gravitational biology and space medicine.

SUMMARY

There are many aspects of cell biology and microgravity that go well beyond this necessarily brief assessment. Although there are cells that directly sense gravitational vectors, the primary tenet presented in this paper is that reduced gravity effects go far beyond the potential role of specialized cells such as plant amyloplasts that aid positioning for plant development in the terrestrial environment. Even though cells may not be able to sense "up" and "down," there are numerous lines of evidence that suggest cells and tissues, for numerous indirect reasons, can be significantly influenced by the microgravity environment. The fact that the influences may be indirect do not detract from their potential importance. The Ex-Me-Cy complex, with its large size and relative instability, is a subcellular structure that could be particularly affected in such an indirect manner. In turn, perturbations of the Ex-Me-Cy complex could influence a significant number of cellular activities and subsequent tissue properties of both plants and animals.

The answers to many of the critical questions in space life sciences, however, will not be solved until more frequent access to the microgravity environment is available, the duration of experimental protocols can be extended to more reasonable lengths, and the proper gravitational force controls are available for experimenters. In this sense, Space Station Freedom is not only important to the future of space life sciences and the biomedical sciences, it is essential.

Acknowledgements — Supported by grants NAGW-1197 and NAGW-2328 from NASA.

REFERENCES

Baldwin, K.M., Herrick, R.E., Ilyina-Kakueva, E., and Oganov, V.S. 1990. Effects of zero gravity on myofibril content and isomyosin distribution in rodent skeletal muscle. *FASEB Journal* 4: 79-83.

Chapes, S.K., Spooner, B.S., Guikema, J.A., and Morrison, D. 1991. Macrophage production of monokines in space. *ASGSB Bulletin* 5(1): 31.

Cipriano, L.F. 1990. An overlooked gravity sensing mechanism. *Physiologist* 34: 72-75.

Cogoli, A., Cogoli, M., Bechler, B., Lorenzi, G., and Gmünder, F. 1990. Microgravity and Mammalian Cells. In: *Microgravity as a Tool in Developmental Biology* (Guyenne, T.D., Ed.). Paris: European Space Agency, p. 11-19. (ESA SP-1123)

Cogoli, A. and Tschopp, A. 1985. Lymphocyte reactivity during spaceflight. *Immunology Today* 6(1): 1-4.

DeBell, L., Guikema, J.A., Wong, P., and Spooner, B.S. 1990. *Rhizobium trifolii* binds to root hairs of white clover seedlings in microgravity. *ASGSB Bulletin* 4(1): 71.

Edidin, M. 1987. Membrane structure and function. In: *Control of Animal Cell Proliferation*, Vol. II (Boynton, A.L. and Leffert, H.L., Eds.). New York: Academic Press, p. 3-36.

Edson, G.D., Fattaey, H.K., and Johnson, T.C. 1991. Cell cycle arrest and cellular differentiation mediated by a cell surface sialoglycopeptide. *Life Sciences* 48: 1813-1820.

Fattaey, H.K., Bascom, C.C., and Johnson, T.C. 1991. Modulation of growth-related gene expression and cell cycle synchronization by a sialoglycopeptide inhibitor. *Experimental Cell Research* 194: 62-68.

Fleming, S.D., Edelman, L.S., and Chapes, S.K. 1991. Effects of corticosterone and microgravity on inflammatory cell production of superoxide. *Journal of Leukocyte Biology* 50: 69-76.

Gould, C.L., Lyte, M., Williams, J., Mandel, A.D., and Sonnenfeld, G. 1987. Inhibited interferon-γ but normal interleukin-3 production from rats flown on the space shuttle. *Aviation, Space, and Environmental Medicine* 58: 983-986.

Gruener, R. 1991. Vector-free gravity interferes with synapse formation. In: *1989-90 NASA Space Biology Accomplishments* (Halstead, T.W., Ed.). Washington, DC: NASA Headquarters, p. 112-116. (NASA TM-4258)

Halstead, T.W. and Dutcher, F.R. 1987. Plants in space. *Annual Review of Plant Physiology* 38: 317-345.

Johnson, T.C. and Sharifi, B.G. 1989. Abrogation of the mitogenic activity of bombesin by a cell surface sialoglycopeptide growth inhibitor. *Biochemical and Biophysical Research Communications* 161: 468-474.

Krikorian, A.D. and Levine, H.G. 1991. Development and growth in space. In: *Plant Physiology*, Vol. X. New York: Academic Press, p. 491-555.

Lewis, M.L. and Hughes-Fulford, M. 1993, in press. Cellular responses to microgravity. In:

Fundamentals of Space Life Sciences. Cambridge, MA: MIT Press.

Life Sciences Division Working Group, SSAAC. 1991. *Cell Biology Discipline Plan.* Washington, DC: NASA Headquarters.

Moos, P.J., Hayes, J.W., Stodiek, L.S., and Luttges, M.W. 1990. Macromolecular assemblies in reduced gravity environments. (American Institute of Aeronautics and Astronautics Paper 90-0027)

Pardee, A.B. 1989. G_1 events and regulation of cell proliferation. *Science* 246: 603-608.

Philpott, D.E., Popova, A., Kato, K., Stevenson, J., Miquel, J., and Sapp, W. 1990. Morphological and biochemical examination of Cosmos 1887 rat heart tissue: Part I - Ultrastructure. *FASEB Journal* 4: 73-78.

Rijken, P.J., de Groot, R.P., Kruijer, W., Boonstra, J., Verkleij, A.J., and de Laat, S.W. 1990. Epidermal growth factor (EGF)-induced signal transduction in A431 cells is sensitive to microgravity. In: *Microgravity as a Tool in Developmental Biology* (Guyenne, T.D., Ed.). Paris: European Space Agency, p. 21-28. (ESA SP-1123)

Sonnenfeld, G. 1989. Response of lymphocytes to a mitogenic stimulus during spaceflight. In: *Cells in Space* (Sibonga, J.D., Mains, R.C., Fast, T.N., Callahan, P.X., and Winget, C.M., Eds.). Moffett Field, CA: NASA, Ames Research Center, p. 77-85. (NASA CP-10034)

Space Studies Board. 1991. Developmental and Cell Biology. In: *Assessment of Programs in Space Biology and Medicine 1991.* Washington, DC: National Academy Press, p. 40-46.

Spooner, B.S. 1992. Gravitational studies in cellular and developmental biology. *Transactions, Kansas Academy of Sciences* 95(1-2): 4-10.

Taylor, G.R., Neale, L.S., and Dardano, J.R. 1986. Immunological analyses of U.S. space shuttle crewmembers. *Aviation, Space, and Environmental Medicine* 57(3): 213-217.

Todd, P. 1989. Gravity-dependent phenomena at the scale of the single cell. *ASGSB Bulletin* 2: 95-113.

Toole-Simms, W.E., Loder, D.K., Fattaey, H.K., and Johnson, T.C. 1991. Effects of a sialoglycopeptide on early events associated with signal transduction. *Journal of Cellular Physiology* 147: 292-297.

Urban, J.E. 1991. Microgravity enhances binding of a bacterioid-inducing molecule in the nitrogen-fixing bacterium *Rhizobium trifolii* (Abstract). *ASGSB Bulletin* 5(1): 83.

Microgravity Research in Plant Biological Systems: Realizing the Potential of Molecular Biology

Norman G. Lewis

Institute of Biological Chemistry, Washington State University, Pullman, WA 99164

Clarence A. Ryan

Institute of Biological Chemistry, Washington State University, Pullman, WA 99164

ABSTRACT

The sole all-pervasive feature of the environment that has helped shape, through evolution, all life on Earth is gravity. The near weightlessness of the Space Station Freedom space environment allows gravitational effects to be essentially uncoupled, thus providing an unprecedented opportunity to manipulate, systematically dissect, study, and exploit the role of gravity in the growth and development of all life forms. New and exciting opportunities are now available to utilize molecular biological and biochemical approaches to study the effects of microgravity on living organisms. By careful experimentation, we can determine how gravity perception occurs, how the resulting signals are produced and transduced, and how or if tissue-specific differences in gene expression occur. Microgravity research can provide unique new approaches to further our basic understanding of development and metabolic processes of cells and organisms, and to further the application of this new knowledge for the betterment of humankind.

INTRODUCTION

Space Station Freedom (SSF) symbolizes a renaissance of NASA's goal to address fundamental questions pertaining to the effect of gravity on living organisms. Many gravitational effects, at least in a phenomenological sense, are already known or suspected, and hence should be amenable to scientific inquiry; others undoubtedly await discovery. SSF will be in service for more than 25 years, and will provide sustained access to a stable microgravity environment, which cannot be duplicated on Earth. Life science experiments on SSF will permit a systematic dissection and analysis, at the molecular and biochemical levels, of various biological phenomena (primarily developmental) that are apparently perturbed in the microgravity environment. Such studies will employ both cell and whole organisms, using all forms of life. It is anticipated that the investigations on SSF will not only benefit space biology, but will also provide novel fundamental and needed knowledge for application to a broad spectrum of human needs. Thus, there exists an unusual opportunity to assess the effects of microgravity and other effects unique to spaceflight on biological processes — in particular, on plants, which display several pronounced gravitropic responses during their life cycles under normal gravitational conditions.

Plant life forms respond to gravitational influences (at 1 g) as demonstrated by gravitropic phenomena. For example, leaning herbaceous plants regain an upright position by increased longitudinal growth on the underside of the stem. Woody gymnosperm and angiosperm plants, on the other hand, restore vertical alignment via altered stem growth patterns resulting in the formation of compression (Timell, 1986; Fengel and Wegener, 1984) and tension (Fengel and Wegener, 1984) wood tissues, respectively, i.e., so-called reaction wood. Gravitropic effects displayed by roots are apparently correlated with the displacement of statoliths in the root tips (Krikorian and Levine, 1991; Volkmann et al., 1991). Many examples of gravitational effects in plants have been described, and some of these phenomena have already been preliminarily studied in space (Halstead and Dutcher, 1987; Halstead et al., 1991). Microgravity experiments with plants ranging from unicellular algae to angiosperms have revealed differences in growth and development when compared with 1 g controls, particularly at the subcellular, cellular, and tissue/organ levels (Halstead and Dutcher, 1987; Halstead et al., 1991). These studies have revealed various phenomenological observations, including: alterations in endoplasmic reticula and ribosomes, "swollen" mitochondria, changes in morphology of the cisternae of dictyosomes, random distribution of amyloplasts (with smaller starch grains), multiple nuclei, chromosomal aberrations, reduction or (partial) inhibition of cell mitosis, disturbances in the mitotic spindle mechanism, differences in cell size and shape, diminution of cellular aggregation capability, alteration in rate(s) of differentiation presumed to lead to more rapid aging, thinner cells walls (with apparently altered biopolymer composition and architecture), disoriented roots (growing upwards rather than downwards), and substantial differences in essential element composition. The inescapable conclusion is that microgravity has a profound effect on plant growth and development.

Phenomenal advances in several areas of plant science have occurred within the last two decades. Knowledge of the molecular biology, biochemistry, physiology, and cell biology of plants has entered a new era in which gene transfer technology has contributed to both fundamental knowledge of plants and to the application of this knowledge to agriculture and related industries. The incorporation of SSF opportunities into this new plant biology can provide an added dimension for interdisciplinary research on plants to answer fundamentally important questions heretofore not possible to address. The following discussions suggest some selected topics in which microgravity research might be the focus of interdisciplinary efforts to contribute new knowledge in areas of plant biology where gravity has been recognized to play major roles in plant growth and development.

GRAVITY SENSING BY PLANTS

A variety of studies on geotropism strongly suggests that when the normal gravitational vector is displaced, a significant alteration of biochemical events occurs. Striking evidence for this assertion comes from biochemical (Lewis et al.) and chemical (Timell, 1986) analyses of reaction wood tissue cells in angiosperms and gymnosperms, which differ substantially in their biopolymer composition and cell wall assembly mechanisms/architecture when compared with normally growing counterparts. This indicates that there is a distinct gravity-sensing mechanism that is initiated perhaps by a perturbation in (mechanical) stress-fields experienced by the cytoskeleton. There are two possibilities for signal transduction: in one scenario, a signal molecule (or molecules) is (are) generated, and bind to one (or various) specific receptor site(s). In the other case, the changes in the stress-field affect conformational changes to the receptor molecule(s), thereby facilitating "docking" of the messenger molecule(s). In both situations, various biochemical events are amplified or repressed either directly or via modulation of gene expression (i.e., via inducing coordinate expression of multiple genes) to redirect a cascade of biochemical events. Although the entire area of gene/biochemical activation in response to the gravitational stimulus is virtually devoid of knowledge, this is an area that can be readily investigated through SSF activities.

SIGNALLING

The signalling mechanisms that regulate genes involved in plant growth and development, plant defense, and host-parasite interactions are under intensive investigation worldwide. In terms of microgravity research, how the gravitational stimulus is transduced to affect biological processes is of fundamental importance to our understanding and exploitation of plant growth and development. It is noteworthy that when this stimulus is essentially removed, a perturbation of normal growth and development follows. This has been elegantly shown by experiments with pine, oats, and mung bean seedlings in the Space Shuttle, where it was observed that the roots were disoriented in microgravity (Cowles et al., 1989). Many researchers have attempted to explain gravitropism in terms of a free-falling statolith in the cytosol coming to rest on the cytoskeleton surface. But how these collisions are subsequently transduced into modulation of gene expression/biochemical events (and resulting physiological responses) is unknown. It is possible that the statolith interaction with the cytoskeleton results in localized stress gradients or concentrations, as suggested above, and that either a chemical message is released (similar to the polypeptide hormone, systemin, released on insect attack (Pearce et al., 1991)), thereby activating a coordinated gene expression response, or the gradient affects a macromolecular conformational change, thereby facilitating binding of the signal molecule(s) to the receptor(s). Whatever scenario holds, a cascade of distinct, overlapping signalling events follows. Thus, incorporation of a coordinated research program involving SSF, utilizing known methodologies and concepts to study signal transduction leading to gravity-stimulation modulation by gene expression, could provide a novel, fundamental approach to furthering our knowledge of signal transduction in plants.

IDENTIFYING GENES INVOLVED WITH GRAVITY COMPENSATION PROCESSES

Not even the simplest molecular biological experiments have been carried out in outer space to date. This is (primarily) because of our inability to cryogenically store plant tissue (-70° C) in orbit on spacecraft, or to isolate and store and manipulate labile compounds under such conditions. Yet, the descriptive changes reported in preliminary space experiments encompassing various aspects of altered growth and development are so striking that they demand our attention. Given the fact that cryogenic facilities will be placed on SSF, experiments using molecular biological techniques involving plant tissues and organs can now be given a high priority, and changes in gene expression (induction or repression) and/or the causal and ensuing biochemical consequences that are influenced by microgravity can be investigated and determined. The next greatest research challenge and opportunity in space will be to manipulate genes in space, to establish how they are regulated, and to investigate their biochemical consequences.

CELL-CELL RECOGNITION AND ADHESION

There is considerable and growing interest in how single cells eventually differentiate into different organs, first via recognition/adhesion interactions leading to pattern formation and, ultimately, via morphogenesis/differentiation (Siu, 1990; Wilkin and Curtis, 1990). That gravitational effects seem to play an important role in such processes has been concluded from several studies, e.g., *Daucus carota* protoplasts were observed to aggregate poorly in microgravity when compared with their 1 g counterparts (Rasmussen et al., 1990). Gravitational effects on cell-cell adhesion may be a more general phenomenon since poor aggregation in microgravity has also been observed with lymphocytes and red blood cells (Halstead et al., 1991). In plants (van Engelen et al., 1991) and animals, specific cell-recognition molecules are associated on the surface of individual cells targeted for aggregation (pattern formation). Given that this process is adversely affected in microgravity, the regulation and composition of cell surface components become logical targets for microgravity research. Current methodologies seem to be well suited for incorporation into an interdisciplinary project in this area, and should provide important information regarding cell patterning.

CELL WALL SYNTHESIS

Plants have, as their major constituents, the cell wall components, i.e., cellulose, lignins, and hemicelluloses. During normal (1 g) growth and development, the plant produces various cell types with distinctive cell walls that differ in the composition and organization of their macromolecular substituents. (It is this process that distinguishes plant and animal cells.) But the biochemistry, including synthesis, deposition, and degradation processes, of these biopolymers is not fully understood, e.g., it is still unknown how cellulose, nature's most abundant organic polymer, is enzymatically synthesized, or how chain (microfibril) orientation is controlled and altered during cell wall synthesis. In a related matter, we do not understand how coordinate synthesis of lignin and hemicelluloses is regulated during cell

wall assembly (Lewis and Yamamoto, 1990). Equally lacking is an understanding of how primary wall assembly and expansion occur, or even how different cell (wall) tissue types are induced or controlled.

Experiments in space to this point (albeit preliminary) have suggested that biopolymer composition and their organization in the cell walls (i.e., architecture) is substantially perturbed in microgravity (Halstead and Dutcher, 1987; Halstead et al., 1991; Lewis et al.). Since the microgravity environment is free of the gravitational stimulus, it can be postulated that these cell walls represent the simplest architecture possible in the growing/developing plant. Thus, a determination of the factors controlling cell wall formation in microgravity will result in development of new strategies to biotechnologically manipulate cell wall formation and, hence, overall growth and development processes.

THE SPACECRAFT AS A BIOCHEMISTRY/ MOLECULAR BIOLOGY LABORATORY

The last 25 to 30 years of spaceflight research has allowed scientists to begin to recognize the research potential of carrying out experiments where the gravity vector has been removed. Many of these experiments have, as stated earlier, given interesting phenomenological observations, which still await clarification at the biochemical and genetic levels. But spaceflight research has been technologically limited in terms of carrying out the best experiments in space biology. These limitations are apparent even today in the experiments designed for SSF. As recently as 1989, NASA designated several areas for investigation in space, and they reflect the need for developing simple growth parameter conditions before more sophisticated research projects can be undertaken (Johnson et al., 1989). These studies include, for example, optimization of plant nutrient and water supplies and plant holding facilities; the ability to grow multiple generations of organisms; determining the effects of microgravity on gas exchange; the control of development at organ and cellular level in microgravity; and other experiments to establish baselines of capabilities. All were included to determine the limitations of basic growth and development processes in microgravity, and all reflect our inability to carry out even the simplest biochemical and molecular biological experiments.

Thus, given the very short time frame to the launch of SSF, the research programs to be included in the SSF agenda must be selected in the immediate future to ensure that the *most modern, highest quality science* is conducted. We recommend that the following be undertaken: (1) rapidly define and design all basic experimental equipment needed for conducting space biology experiments, which is currently under way; and (2) define 8-12 key experiments (or questions) that need to be answered in microgravity for each discipline, and identify and assemble *teams* of investigators (inter- and multidisciplinary) who have the ability to use today's technologies and today's ideas to address segments thereof. This will ensure that the best science will be undertaken and completed and that the potential of microgravity will be realized.

CONCLUDING REMARKS

The availability of SSF provides an unprecedented and exciting opportunity to systematically determine how gravity affects the growth and development of all life forms. Although a particular emphasis was placed, in the preceding sections, upon plant systems that show fairly unique gravitropic responses, fascinating differences are also noted with mammalian systems (and other organisms) in microgravity. Hence, molecular biological and biochemical studies can be anticipated to yield important information on a variety of subject areas, such as bone formation and structure, immunology, muscle formation, and the cardiovascular system.

This paper focussed on plants, which represent our principal source of food, clothing, shelter, and medicinal compounds. A systematic examination of the effects of gravity on plant growth and development in the absence of gravity, at the genetic/biochemical level, will allow us to identify and design new ways to biotechnologically exploit plant life in a manner hitherto not possible. It can be anticipated that this will greatly assist in resolving numerous outstanding technical questions, including finding better ways to produce foodstuffs, enhancing the production of medicinals, and improving the supply and quality of wood and related fibrous materials for future generations.

REFERENCES

Cowles, J.R., LeMay, R., Jahns, G., Scheld, W.H., and Peterson, C. 1989. Lignification in young plant seedlings grown on Earth and aboard the Space Shuttle. *ACS Symposium Series* 399: 203-213.

Fengel, D. and Wegener, G. 1984. *Wood: Chemistry, Ultrastructure, Reactions.* Berlin: Walter de Gruyter.

Halstead, T.W. and Dutcher, F.R. 1987. Plants in space. *Annual Review of Plant Physiology* 38: 317-345 (and references therein).

Halstead, T.W., Todd, P., and Powers, J.V. (Eds.) 1991. Gravity and the Cell: Report of a Conference Held December 1-3, 1988 in Washington, D.C. *ASGSB Bulletin* 4(2): 1-260 and references therein.

Johnson, C.C., Arno, R.D., and Mains, R. (Eds.) 1989. *Life Science Research Objectives and Representative Experiments for the Space Station.* Washington, DC: NASA, 300 p. (NASA TM-89445)

Krikorian, A.D. and Levine, H.G. 1991. Development and growth in space. In: *Plant Physiology: A Treatise: Growth and Development,* Vol. X (Bidwell, R.G.S. and Steward, F.C., Eds.). New York: Academic Press, p. 491-555 (and references therein).

Lewis, N.G. et al. (unpublished results).

Lewis, N.G. and Yamamoto, E. 1990. Lignin: Occurrence, biogenesis and biodegradation. *Annual Review of Plant Physiology* 41: 455-496.

Pearce, G., Strydom, D., Johnson, S., and Ryan, C.A. 1991. A polypeptide from tomato leaves induces wound-inducible proteinase inhibitor proteins. *Science* 253: 895-898.

Rasmussen, O., Gmünder, F., Tairbekov, M., Kordyum, E.L., Lozovaya, V.V., Baggerud, C., and Iverson, T.H. 1990. Plant protoplast development on "Biokosmos 9." In: *Proceedings of the Fourth European Symposium on Life Sciences Research in Space,* Trieste, Italy, May 28-June 1, 1990. Paris: European Space Agency, p. 527-530. (ESA SP-307)

Siu, C.H. 1990. Cell-cell adhesion molecules in *Dictyostelium. Bioessays* 12(8): 357-362.

Timell, T.E. 1986. *Compression Wood in Gymnosperms,* Vol. 1-3. Berlin: Springer-Verlag, 2150 p.

van Engelen, F.A., Sterk, P., Booij, H., Cordewener, J.H.G., Rook, W., van Kammen, A.B., and de Vries, S.C. 1991. Heterogeneity and cell type-specific localization of a cell wall glycoprotein from carrot suspension cells. *Plant Physiology* 96: 705-712.

Volkmann, D., Buchen, B., Hejnowicz, Z., Tewinkel, M., and Sievers, A. 1991. Oriented movement of statoliths studied in a reduced gravitational field during parabolic flights of rockets. *Planta* 185: 152-161.

Wilkin, G.P. and Curtis, R. 1990. Cell adhesion molecules and ion pumps — Do ion fluxes regulate neuron migration? *Bioessays* 12(6): 287-288.

Life: Origin and Evolution on Earth — How Can We Escape?

CLEMENT L. MARKERT
Department of Animal Science, North Carolina State University, Raleigh, NC 27695

ABRAHAM D. KRIKORIAN
Department of Biochemistry and Cell Biology, State University of New York, Stony Brook, NY 11794

ABSTRACT

Exploitation of gene regulation rather than the creation of new genes has been predominantly responsible for the evolutionary advances in animals and plants that are widely recognized today. Until very recently it was not possible to examine life in the absence of gravity. We can now imagine forms of life in the universe adapting to circumstances different from those found on Earth. Our own life forms would surely become different in time if they were transferred to other planets with different conditions, including much lower or higher gravity.

Life arose on Earth nearly four billion years ago as membrane-contained biochemical and biophysical systems that were isolated from each other and protected from the environment. These systems could enlarge, subdivide, and reproduce as individual organisms. Organic and inorganic molecules were selectively absorbed and processed, and some molecules were then excreted (cf. Schopf, 1983). Thus, a form of biological replication — crude and erratic at first — slowly evolved, stabilized, and began to exhibit characteristics appropriate for the survival of the fittest. It took a billion years or more for this system to evolve beyond a bacterium-like cell stage. The basic biochemical patterns and the mechanisms for replicating the biochemical architecture with integrity must have evolved very slowly during an additional two billion years. Perhaps another 700 million years passed during the evolution of multicellular organisms, which takes us to the middle of the Cambrian era of evolutionary development (cf. Avers, 1989). Then, within the next 500 million years, all of the contemporary divisions of plant and animal phyla developed. Mammals and higher plants have been around for about 200 million years, since the Triassic age of the Mesozoic era, but in the case of the mammals at least, these organisms were primitive and very small and not numerous for the next 140 million years. Only since the dinosaurs disappeared at the beginning of the Cretaceous era have mammals and flowering plants undergone an explosive evolution. The first primates go back about 60 million years.

Contemporary humans shared a common ancestor with our close relatives, the great apes, about 10 millions years ago, and the ancestors of contemporary human beings, *Homo sapiens*, have been recognizable for perhaps the last two or three hundred thousand years. Modern humans biologically indistinguishable from ourselves may go back only 40 or 50 thousand years. Thus, we are very late comers to Earth.

The extraordinarily rapid development that we seem to see in the formation of the most highly evolved organisms today in terms of their size, mobility, and capacity to dominate the environment reflects perhaps an accelerating rate of diversification of life forms. This diversity is based upon a reassortment and rearrangement of basically the same building blocks (cells) — building blocks that were created by the billions of years of early evolution (cf. Briggs and Crowther, 1990). Of course, human cultural, scientific, and technological evolution that gave rise to contemporary civilizations began just a few thousand years ago, and most developments have occurred within the last few hundred years. This recitation of freshman biology has been made to emphasize that the exploitation of gene regulation rather than the creation of new genes is primarily responsible for the evolutionary advances in metaphytes and metazoans — i.e., multicellular plants and animals. In animals especially, but plants as well, the precise movement and/or associations of cells and their parts, all sensitive to gravity (more precisely, in plants both meristematic and non-meristematic regions may be graviresponsive), are critical for recent evolutionary developments.

All of this evolution, of course, proceeded with gravity as an ever present and constant aspect of the environment. Biologists have investigated many of the specific mechanisms of evolution, as well as the basic biochemical and biophysical nature of life itself, but we have given practically no attention to the role that gravity has played and is playing in the structure and functioning of contemporary organisms (cf. Stebbins, 1982; Avers, 1989). The explanation is

35

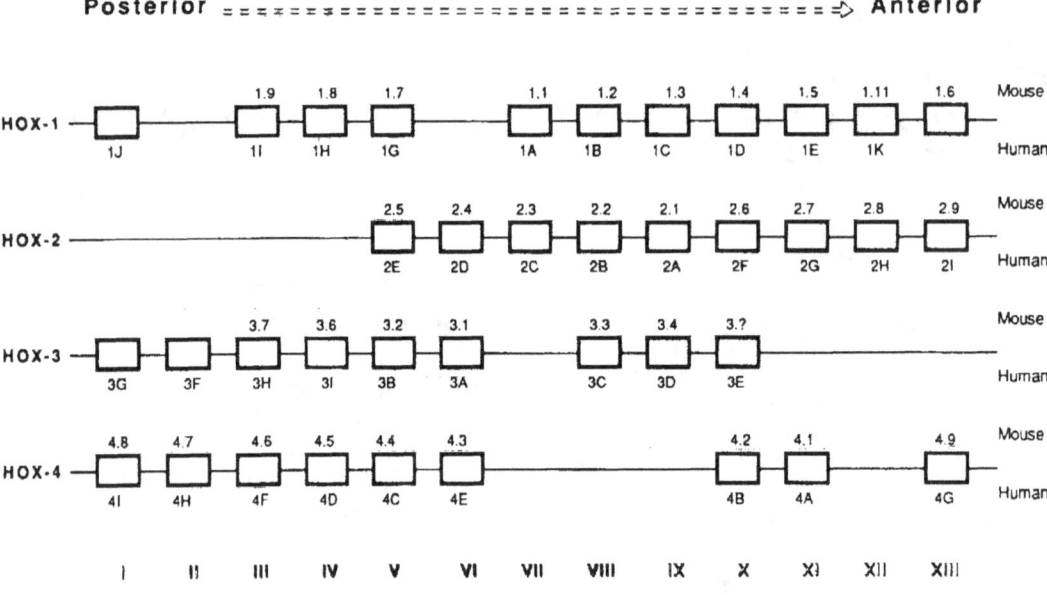

Figure 1. Distribution along chromosomes of homologous genes in mouse and humans that contain homeoboxes. These genes are activated in the embryo in a temporal and spatial sequence that corresponds to the linear position of the genes along the chromosome. The homology of the homeobox sequences among many taxa has been highly conserved throughout evolution. Diagram redrawn from Murtha et al., 1991.

simple, of course. Until very recently, it was not possible to examine life in the absence of gravity. Now we can do so with the space vehicles and stations and with our ability to engage in space travel (cf. Krikorian and Levine, 1991).

Life itself has made the Earth a very different place from what it was before life evolved. There is a film of living material covering virtually all of the land surface of the Earth and, of course, also in the water that covers most of the Earth's surface. The gaseous atmosphere has been changed by the activity of life, and thus has changed the environment with consequent effects on the further evolution of organisms. Life continuously creates a changing environment and then responds to the changes by further evolution (cf. Briggs and Crowther, 1990). We have never escaped from gravity except momentarily and then only in recent years. It is possible to imagine many other forms of life in the universe evolving and adapting to quite different circumstances from those we find on Earth, and our own life forms on Earth would surely become different if they were transferred to other planets with different physical conditions, including much lower or much higher gravity.

The current importance of understanding the role of gravity in our own evolution and development stems primarily from the fact that we plan to send explorers to the moon and to Mars and perhaps elsewhere and to take other forms of life with us (cf. e.g., Robbins Committee Report, 1988). The consequences for the structure, physiology, and development of ourselves and other organisms will surely be profound, and it behooves us to understand the significance of gravity in determining our own basic biological nature before we meet these extraterrestrial challenges. At the most basic level of life, the individual cell, we find a vast array of molecules, of cell organelles, and of various elaborate structures that collectively make possible the biochemical activities that keep the cell alive, developing, and reproducing (cf. Alberts et al., 1989).

Recently, we have come to recognize the organization of specific groups of genes containing sequences (homeoboxes) that seem to specify in time and place the development of the organism (cf. e.g., Murtha et al., 1991). In mice and humans, both species have been extensively investigated; four different clusters of homeoboxes have been identified, each on a different chromosome. Within each of these clusters, there are about ten genes distributed along the chromosome in a precise order that is the same for mice and humans. This physical arrange-

LIFE AND EVOLUTION WITHOUT GRAVITY

Parascaris equorum

Normal Development

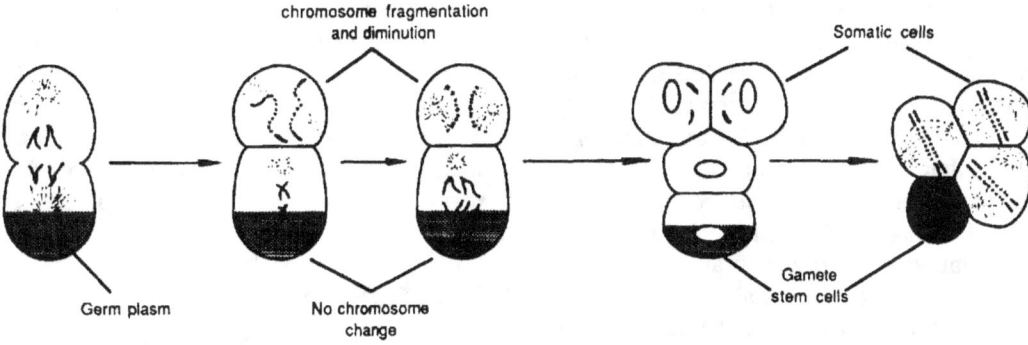

Abnormal Development After Displacement of Germ Plasm by Centrifugation

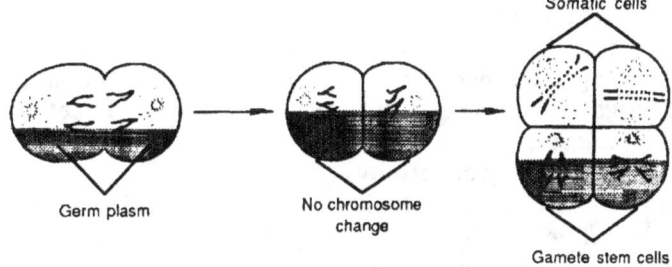

Figure 2. Diagram showing results of centrifugal displacement of germinal plasm in *Parascaris [Ascaris megalocephala]*. Presence of germinal plasm in a cell prevents chromosomal diminution and fragmentation of large chromosomes into many small chromosomes. The presence of intact large chromosomes are essential for germ cell development. Cells with reduced chromosomal content become somatic cells. After Boveri, 1910.

ment — the anterior-posterior sequence — corresponds to the time of activation of these genes during the course of early development (Figure 1). Similar patterns of homeobox organization have been discovered in a wide variety of organisms stretching all the way back to the most primitive multicellular organisms. The fact that these genes have been extraordinarily well conserved during the course of evolution indicates that they must have a fundamental role in determining the structure of organisms. All of the evidence we have so far indicates that they do play such roles in developing the basic morphology, very likely by affecting cell associations and relative rates of multiplication that could define the axial and other structures of the developing organism. Clearly, these genes could not be abnormally activated or inactivated without drastic effects on the development of the organism. Progressive slight changes, in accord with what we might expect in evolution, could of course lead to the variety of morphologies that we see in different organisms. Obviously, however, the slightest per-

turbation in the activation of these genes during the course of development would have profound consequences. Thus, the role of gravity in perturbing the program of activation of these homeobox genes must be understood in detail if we are to safely reproduce in space. Molecules and organelles can be moved within the cell by centrifugation and, therefore, must respond to gravitational forces exerted on the cell; such redistributions of components within a cell could prove fatal or at least could change the development of embryos. The temporary redistribution of cell contents by centrifugation can frequently be overcome and the original state of the cell restored. Many forces are involved in molecular traffic to move and orient molecules and organelles within the cell. Gravity is certainly one of them (cf. Halstead et al., 1991). How the cell manages without gravity and how it changes in the absence of gravity are basic questions that only prolonged life on a space station will enable us to answer (cf. Souza and Halstead, 1985; Asashima and Malacinski, 1990).

We know from the experience acquired so far on space vehicles and stations, our own and those of the Soviet astronauts (cf. Garshnek, 1988), that profound deleterious physiological effects do occur and jeopardize the functional capacity and even the survival of human beings. To counteract or circumvent these effects, we need to know more of the basic biochemistry and biophysics of the cell and of the whole organism in conditions of reduced gravity. Such knowledge is essential in order to make space travel and residence on the moon or Mars practicable. Reproduction among highly differentiated organisms such as mammals, birds, and other vertebrates and probably also invertebrates will obviously be seriously affected by the absence of gravity (cf. Guyenne, 1990). Organisms with large amounts of yolk in their eggs, such as amphibians and birds, will be seriously affected by the redistribution of components within the egg when gravity is greatly reduced. Homolecithal eggs, such as those of human beings, would probably be less affected, but even in these eggs, the molecular and organellar traffic would surely be affected by the prolonged absence of gravity. The basic hereditary organelle of the cell, the chromosome, and the many molecules with which it interacts to produce precise patterns of gene activity during development and in normal adult physiological function would also be influenced by gravity (Figure 2). Can such animals and plants reproduce on a space platform and through several successive generations? Only research on the station could answer these questions. Moreover, we do not now have the insights required to ask many questions that will surely become obvious by experience in space. Stable long term experiments through several generations are needed on a space station in order to know our capabilities and to protect them for the long trips to Mars or residence on the moon or Mars. Once we develop a clear understanding of the consequences of the absence of gravity on our physiological and developmental capacities, such information will also enlighten our response to challenges here on Earth. In fact, such information and understanding would be of great value even if we never ventured into space.

In summary, there are two major reasons for building a space station and carrying on long-term experiments in biological and biomedical sciences on that platform. First, we must do so if we are to keep human beings and other organisms in space or on other planets, the moon, or Mars for extended periods of time. We cannot survive there under present circumstances, and we do not know enough to overcome the hostile environments due to the absence of gravity and to the various other challenges of space, such as radiation (McCormack et al., 1989). Second, we will achieve deeper insight into the nature of our own biological structures and activities by understanding the significance of gravity in the development of morphology and physiological function (cf. e.g., Oser and Battrick, 1989). We cannot predict what the research on organisms in the absence of gravity will produce. Otherwise, we would not need to do the research. But that there will be significant enlightenment seems obvious. Surprises there will be, and we should be enthusiastic in welcoming the knowledge and insights that will surely result from biomedical research in space.

REFERENCES

Alberts, B., Bray, D., Lewis, J., Raff, M., Roberts, K., and Watson, J.D. 1989. *Molecular Biology of the Cell*, Second Edition. New York: Garland Publishing.

Asashima, M. and Malacinski, G.M. (Eds.) 1990. *Fundamentals of Space Biology*. Tokyo/Berlin: Japan Scientific Societies Press/Springer-Verlag, 203 p.

Avers, C. 1989. *Process & Pattern in Evolution*. New York: Oxford University Press.

Boveri, T. 1910. Ueber die Teilung centrifugierter Eier von *Ascaris megalocephala*. *Archiv der Entwicklungsmechanik und Organismen* 30: 101-125.

Briggs, D.E.G. and Crowther, P.R. (Eds.) 1990. *Paleobiology: A Synthesis*. Oxford: Blackwell.

Garshnek, V. 1988. Soviet space flight: The human element. *ASGSB Bulletin* 1: 67-80.

Guyenne, T.D. (Ed.) 1990. *Microgravity as a Tool in Developmental Biology*. Paris: European Space Agency. (ESA SP-1123) (ISBN 92-9092-064-5)

Halstead, T.W., Todd, P., and Powers, J.V. (Eds.) 1991. Gravity and the Cell: Report of a Conference Held December 1-3, 1988 in Washington, D.C. *ASGSB Bulletin* 4(2): 1-260.

Krikorian, A.D. and Levine, H.G. 1991. Development and growth in space, Chapter 8. In: *Plant Physiology: A Treatise*, Vol. X (Bidwell, R.G.S., Ed.). Orlando, FL: Academic Press, p. 491-555.

McCormack, P.E., Swenberg, C.E., and Buecker, H. (Eds.) 1989. *Terrestrial and Space Radiation and its Biological Effects*. NATO Advanced Study Institute

Series A. Life Sciences. 154. Corfu, Greece, 11-25 October 1988. New York: Plenum.

Murtha, M.T., Leckman, J.F., and Ruddle, F.H. 1991. Detection of homeobox genes in development and evolution. *Proceedings of the National Academy of Sciences, USA* 88: 10711-10715.

Oser, H. and Battrick, B. (Eds.) 1989. *Life Sciences Research in Space*. Paris: European Space Agency, 141 p. (ESA SP-1105) (ISSN 0379-6566)

Robbins Committee Report. 1988. *Exploring the Living Universe: A Strategy for Space Life Sciences. A Report of the NASA Life Sciences Strategic Planning Study Committee*. Washington, DC: NASA, 231 p.

Schopf, J.W. (Ed.) 1983. *Earth's Earliest Biosphere. Its Origin and Evolution*. Princeton, NJ: Princeton University Press, 543 p.

Souza, K.A. and Halstead, T.W. (Eds.) 1985. *NASA Developmental Biology Workshop*, Arlington, Virginia, May 1984. Moffett Field, CA: NASA, Ames Research Center, 92 p. (NASA TM-86756)

Stebbins, G.L. 1982. *Darwin to DNA, Molecules to Humanity*. San Francisco, CA: W.H. Freeman.